DOMAIN DECOMPOSITION METHODS
FOR NONCONFORMING FINITE ELEMENT DISCRETIZATIONS

DOMAIN DECOMPOSITION METHODS
FOR NONCONFORMING FINITE ELEMENT DISCRETIZATIONS

GU JINSHENG

Nova Science Publishers, Inc.
Commack, New York

Editorial Production: Susan Boriotti
Office Manager: Annette Hellinger
Graphics: Frank Grucci and John T'Lustachowski
Information Editor: Tatiana Shohov
Book Production: Donna Dennis, Patrick Davin, Christine Mathosian and Tammy Sauter
Circulation: Maryanne Schmidt
Marketing/Sales: Cathy DeGregory

Library of Congress Cataloging-in-Publication Data
available upon request

ISBN 1-56072-614-8

6080 Jericho Turnpike, Suite 207
Commack, New York 11725
Tele. 516-499-3103 Fax 516-499-3146
e-mail: Novascience@earthlink.net
e-mail: Novascil@aol.com
Web Site: http://www.nexusworld.com/nova

Printed in the United States of America

Contents

Part II Domain Decomposition Methods for Elliptic Problems of 4th Order

Preface

Domain decomposition refers to numerical methods for obtaining solutions of scientific and engineering problems by combining solutions to problems posed on physical subdomains, or, more generally, by combining solutions to appropriately constructed subproblems. It has been a subject of intense interest recently because of its suitability for implementation on high performance computer architectures. It is well known that the nonconforming finite elements are widely used in and effective for the solving of partial differential equations derived from mechanics and engineering, because they have fewer degrees of freedom, simpler basis functions and better convergence behavior. But, there has been no extensive study of domain decomposition methods with nonconforming finite elements which lack the global continuity. Therefore, a rather systematic investigation on domain decomposition methods with nonconforming elements is of great significance and this is what the present book tries to achieve. The theoretical breakthrough is the establishment of a series of essential estimates, especially the extension theorems for nonconforming elements, which play key roles in domain decomposition analysis. There are also many originalities in the design of the domain decomposition algorithms for the nonconforming finite element discretizations, according to the features of the nonconforming elements. The existing domain decomposition methods developed in the conforming finite element discrete case can be revised properly and extended to the nonconforming finite element discrete case correspondingly. These algorithms, nonoverlap or overlap, are as efficient as their counterparts in the conforming cases, and even easier in implementation.

With the domain decomposition methods for nonconforming finite element discretizations of elliptic problems as the main content, this book is intended to graduate students, researchers and engineers working on one of the subjects: numerical analysis, large scale scientific/engineering computing, and parallel algorithms.

I am indebted to Prof. Widlund O.B., Prof. Lin Qun, Prof. Yu Dehao, and Prof. Wang Lie–heng for their interests and many suggestions on my research. I am especially grateful to my graduate advisor, Prof. Hu Xiancheng for his long time direction, support, encouragement and fruitful cooperation. Special thanks go to Dr. Chu Delin, Dr. Zhang Sheng, Dr. Huang Jianguo and Dr. Xie Zhenghui for many inspiring discussions and beneficial advice. I must thank my wife, Jiang Shuqin, for her patience, understanding and assistance in looking after our child and proofreading the manuscript. Finally, I express my sincere gratitude to Frank Columbus, President and Editor–in–Chief, Nadya S. Columbus, Vice–President, and their colleagues of Nova Science Publishers, Inc., for their consistent help to make the manuscript published as soon as possible.

The author welcomes in advance all comments, suggestions, criticisms, etc.

Jinsheng Gu
Department of Mathematics
Capital Normal University
Beijing, P.R. China

June, 1998

Chapter 1

Introduction

1.1 Overview

Many mathematical models of scientific and engineering problems are described by partial differential equations. The numerical solution of these problems of mathematical physics is always the central issue of scientific and engineering computing. With the development of parallel computer, domain decomposition methods have attracted more and more attention in the area of large scale scientific computation. The domain is decomposed into subdomains, overlap or nonoverlap. The overlap (resp. nonoverlap) domain decomposition methods are often referred to as the Schwarz alternating methods (resp. the substructuring methods). In each iteration step, a number of smaller subproblems, which correspond to the restriction of the original problem to the subdomains, or to a problem with a coarse mesh, are solved. The number of subproblems can be large and these methods are well suited for parallel computer architecture. The domain decomposition methods have become an important focus in the field of computational mathematics, because they have the following additional advantages[36]:

1. Large scale global problems can be divided into small scale independent subproblems which can be solved simultaneously;

2. New methods can be developed to solve the algebraic equations from the stage of mathematical model, not just only from the stage of matrix analysis;

3. The fast algorithms, such as the fast Fourier transformation (FFT), the spectral method, and the Tau method, can be applied for the subproblems on the regular subdomains;

4. The global mesh does not need to be quasi-uniform, especially different discrete schemes can be used to solve different subproblems. This is of extremely importance for problems on very complicated domains, such as the boiler combustion problem;

5. Different mathematical models can be adopted for different subdomains, such that it could be more accurate to simulate the practical problems, such as the transonic problem.

Up to now, the domain decomposition methods can be classified mainly into three types: overlap, nonoverlap and multilevel, which are surveyed as below:

- The concept of overlap domain decomposition[76] was first proposed by H.A. Schwarz to study the regularity properties of the soluton of an elliptic boundary value problem on irrugular domain in 1869. It involves solving problems on two subdomains sequentially. Its convergence was established using maximum principle. Later, it was extended to the partial differential equations of linear elasticity in 1930's. The domain decomposition method based on this concept is therefore called the Schwarz alternating method. Lions[61] interpreted the method in a variational framework. Many subdomain cases were also considered. Because of the sequential nature of the method, it is not ideal for parallel computations. Dryja[32] introduced the additive Schwarz method to remove the sequential behavior and make it possible to solve all the subproblems in parallel. The additive Schwarz method can be regarded as an iterative method for an equivalent system with a better conditioner number. The optimal convergence properties of the algorithm for the second order problems discretized by the piecewise linear conforming elements were proven. Generalization to certain stationary and parabolic convection–diffusion problems was carried out by Cai[20]. A comparison of the overlap and nonoverlap domain decomposition methods was presented in [58].

- The idea of nonoverlap domain decomposition or substructuring has a long history from the structural analysis community[75]. The iterative substructuring methods for the second order problems discretized by the piecewise linear conforming elements were developed by Bramble[11–15], Bjørstad[8], Dryja[31,32] and Widlund[83,84]. It is rather complicated in the case of multi–subdomains where the internal cross points exist. A cross point is defined to be the common boundary point of more than two subdomains. In the theoretical analysis of the substructuring methods, the so–called extension theorem plays a key role. Techniques for analyzing some iterative substructuring methods using the additive Schwarz framework were developed, see e.g.[8]. It is demonstrated that certain iterative substructuring methods can be viewed as additive Schwarz methods with a set of special overlap subdomains. The relationship between the two type schemes was also discussed in details. The algorithms and results have been successively extended to spectral and mortar finite element discrete cases in [85].

- Multilevel methods, such as multigrid methods, are among the most efficient methods for linear equations arising from elliptic problems[17,52]. Recently, with the increasing interest in parallel computation, several new multilevel methods have been developed, see, e.g., [7,16,66,78,88] and the references therein.

It is well known that the nonconforming finite elements are widely used in and effective for the solving of partial differential equations derived from mechanics and engineering[26,79], because they have fewer degrees of freedom, simpler basis functions and better convergence behavior. For example, the Morley nonconforming elements[67] have more advantages than the conforming elements such as the Argyris elements[2] for the solving of the plate bending problems. But, there has been no extensive study of domain decomposition methods with nonconforming finite elements which lack the global continuity. Therefore, a rather systematic investigation on domain decomposition methods with nonconforming elements is of great significance and this is the aim of the present book. The theoretical breakthrough is the establishment of the extension theorems for nonconforming elements. There are also many originalities in the design of the domain decomposition algorithms for the nonconforming finite element discretizations, according to the characteristics of the nonconforming elements.

From [12] and [84], we know that the extension theorem is of critical importance in the domain decomposition analysis, especially in the nonoverlap domain decomposition analysis. We also know that the extension theorem holds for conforming elements, hence the domain decomposition analysis can be performed for the conforming finite element discrete problems. When domain decomposition methods with nonconforming elements are studied, a core question in domain decompisition analysis is "Do the extension theorems hold for nonconforming finite elements?"

For this purpose, we have established the extension theorems for nonconforming elements, based on the conforming interpolation operator and the further error estimates of the nonconforming finite element approximate solution under weak conditions[49,50]. Note that for the second kind nonconforming element discretizations, the values at the internal cross points are unnecessary to be obtained and in fact impossible to be obtained. Therefore, the internal cross points don't need to be handled, and a series of domain decomposition algorithms have been proposed, which are simple and high parallel.

Although the nonconforming finite element function lacks the global continuity, the theoretical foundation can be established and the existing domain decomposition methods developed in the conforming finite element discrete case can be revised properly and extended to the nonconforming finite element discrete case correspondingly. For the second order elliptic problems discretized by the nonconforming element methods, all the domain decomposition algorthms, nonoverlap or overlap, are as efficient as their counterparts in the conforming cases, and even easier in implementation. For the fourth order elliptic problems in the continuous case and in the Morley nonconforming discrete case, the Schwarz alternating algorithms are investigated and plenty of numerical results are given. For the plate bending problem discretized by finite elements, especially by the nonconforming elements, the extension theorems are shown and applied to the analysis of the substructuring algorithms with two subdomains.

The book is arranged as follows. In the remainder of the present chapter, we review some basic Sobolev spaces and lemmas. We also review and classify the nonconforming finite elements for the second order elliptic problems. Since the domain decomposition methods could be regarded as variations of preconditioned conjugate gradient methods (PCG), we present the key step of PCG to end this chapter. Chapter 2 is devoted to the establishment of the theoretical foundation of domain decomposition analysis for the nonconforming finite element discretizations of second order elliptic problems. Based on the theoretical results of Chapter 2, the domain decomposition methods are developed and analyzed from Chapter 3 to Chapter 9 for the second order problems. Chapter 3 presents a brief survey of all existing substructuring algorithms with two subdomains, some of which are discussed in details. The substructuring algorithms with multi–subdomains are designed and analyzed in Chapter n, $4 \leq n \leq 8$. It is rather complicated in this case if there exist the internal cross points. For the solving of the first kind nonconforming element discretizations, the iterative algorithm introduced in [9] is generalized in Chapter 4 and a series of preconditioners are constructed in Chapter 5, with the techniques of the separation of the internal cross points from other mesh nodes. For the solving of the second kind nonconforming element discretizations, two iterative algorithms are developed in the red–black subdivision case in Chapter 6 and in the general subdivision case in Chapter 7, while several preconditioners are proposed in Chapter 8. These methods are simple and high parallel, because the internal cross points do not need to be coped with in this case. Chapter 9 considers the overlap domain decomposition methods. In Sect.9.1, the Schwarz alternating method is developed and analyzed for the Crouzeix–Raviart element discretizations of the nonselfadjoint problems, based on the discrete maximum principle. And in Sect.9.2, with the projection operator theory, this method is stated and analyzed for the generic finite elements, conforming or nonconforming, only if certain conditions are satisfied. The domain decomposition methods for the fourth order elliptic problems are presented in Chapter 10. The main result of this chapter is the extension theorems for nonconforming plate elements, which is shown in Sect.10.1. The substructuring algorithm with two subdomains is also analyzed. Finally in Sect.10.2, the Schwarz alternating method for the continuous problem and for the nonconforming discrete problem is presented and numerical experiments are given also.

1.2 Sobolev spaces

Throughout the book, we use the standard conventions of Sobolev spaces[1,60]. Let $\Omega \subset \Re^d$ $(d \geq 2)$ be a bounded Lipschitz domain and $L^p(\Omega)$ the Banach sapce of the pth power integrable functions.

The *multi–index* notation is employed for partial derivatives. Let $\alpha \equiv (\alpha_1, \cdots, \alpha_d)$ be a nonnegative multi–index representing the order of the partial derivatives, $|\alpha| \equiv \alpha_1 + \cdots + \alpha_d$, and let

$$\partial_\alpha \triangleq \frac{\partial^{|\alpha|}}{\partial x_1^{\alpha_1} \cdots \partial x_d^{\alpha_d}}$$

The Sobolev space $W^{m,p}(\Omega)$ is defined by

$$W^{m,p}(\Omega) \triangleq \{v : \ \partial_\alpha v \in L^p(\Omega), \forall \ |\alpha| \leq m\}$$

with a norm

$$\|v\|_{W^{m,p}(\Omega)} = \Big(\sum_{|\alpha| \le m} \|\partial_\alpha v\|^p_{L^p(\Omega)} \Big)^{\frac{1}{p}}.$$

We also need the seminorm

$$|v|_{W^{m,p}(\Omega)} = \Big(\sum_{|\alpha| = m} \|\partial_\alpha v\|^p_{L^p(\Omega)} \Big)^{\frac{1}{p}}.$$

The most useful space in this book is the case $p = 2$, which by convention is denoted by $H^m(\Omega)$,

$$H^m(\Omega) \triangleq W^{m,2}(\Omega).$$

This is a Hilbert space. For $v \in H^m(\Omega)$, the $H^m(\Omega)$ norm of v is sometimes denoted as $\|v\|_{m,\Omega}$ instead of $\|v\|_{H^m(\Omega)}$ for convenience. We shall also make frequently use of the semi–norm

$$|v|_{m,\Omega} = |v|_{H^m(\Omega)} = \Big(\sum_{|\alpha| = m} \int_\Omega |\partial_\alpha v|^2 dx \Big)^{\frac{1}{2}}.$$

To study the homogeneous Dirichlet problem, we need the Sobolev space

$$H_0^m(\Omega) \triangleq \{v \in H^m(\Omega) : \ \partial_\alpha v|_{\partial\Omega} = 0, \ \forall\ |\alpha| \le m-1\}.$$

This is the closure of the space of $C_0^\infty(\Omega)$ with respect to $H^m(\Omega)$–norm.

Next we summarize several frequently used inequalities, which establish equivalences of certain norms in the subspaces of $H^m(\Omega)$. The following two can be found in [68], which are the famous Poincaré inequalities.

Lemma 1.2.1. There exists a positive constant C, such that

$$\|u\|_{L^2(\Omega)} \le C|u|_{H^1(\Omega)}, \qquad \forall\ u \in H_0^1(\Omega)$$

Lemma 1.2.2. There exists a positive constant C, such that

$$\|u - \frac{\int_\Omega u dx}{meas(\Omega)}\|_{L^2(\Omega)} \le C|u|_{H^1(\Omega)}, \qquad \forall\ u \in H^1(\Omega)$$

The following lemma is the generalized Poincaré inequality, which can be found from [26].

Lemma 1.2.3. There exists a positive constant C, such that

$$\|u\|_{H^s(\Omega)} \le C|u|_{H^m(\Omega)}, \quad \forall\ s \le m, \quad \forall\ u \in H^m(\Omega)$$

Therefore, the semi–norm $|\cdot|_{m,\Omega}$ is a norm over the space $H_0^m(\Omega)$, equivalent to the norm $\|\cdot\|_{m,\Omega}$.

Other classic results related to the Sobolev spaces, such as the trace theorem, the imbedding theorems, the interpolation theory, could be found in many references, see, e.g., [1,26,28,40,60,68,69]. Hence we omit them here.

1.3 Nonconforming elements for second order problems

Suppose that $\Omega_h = \{e\}$ is a quasi-uniform mesh of Ω, i.e., Ω_h satisfies

$$\sup_{e\in\Omega_h} \inf_{B_r\supset e} r \le ch, \qquad \inf_{e\in\Omega_h} \sup_{B_r\subset e} r \ge Ch,$$

where e, a triangle or a quadrilateral, represents the typical element in Ω_h, B_r is a region bounded by the circle of radius r, $h = \max_{e\in\Omega_h} h_e$ is the mesh parameter and $h_e = \inf_{B_r\supset e} r$. Here and later, c and C, with or without subscript, denote generic positive constants independent of h. Associated with Ω_h, we can introduce many finite element spaces [6,26,34]. This book mainly considers the nonconforming finite elements. For second order elliptic problems, five nonconforming elements are described as follows.

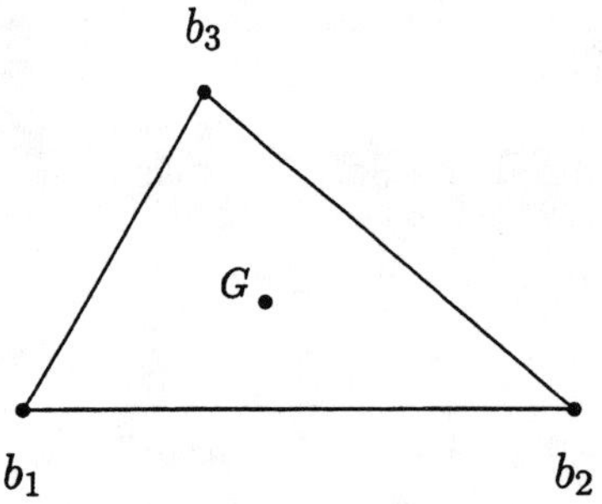

Fig.1.3.1

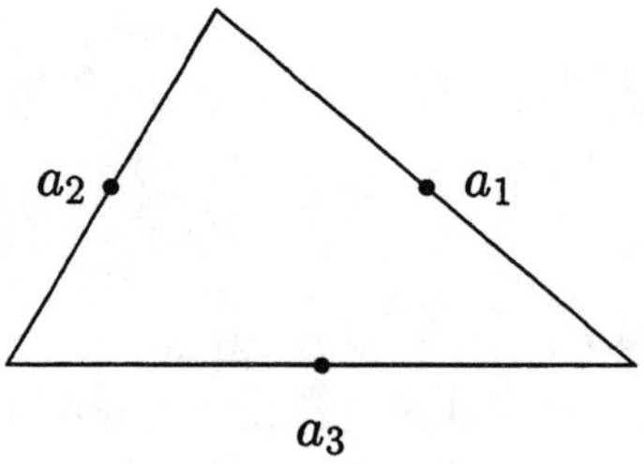

Fig.1.3.2

The Carey element space and the Crouzeix–Raviart element space require every element of Ω_h, the triangulation of Ω, be a triangle. For the following two element spaces, suppose that e is a triangle with vertices b_i $(i = 1, 2, 3)$. Let λ_i (resp. a_i) label the area coordinate (resp. opposite edge midpoint) corresponding to b_i.

1.3.1 Carey element

The Carey element space[21] on e, denoted $\mathcal{C}_e$, is spanned by λ_1, λ_2, λ_3 and $\lambda_1\lambda_2 + \lambda_2\lambda_3 + \lambda_3\lambda_1$, which is the basis at G, the barycenter of e; see Fig.1.3.1. Here, b_1, b_2, b_3, and G are interpolation points. The shape function for e is defined by

$$v = \sum_{i=1}^{3} v(b_i)\lambda_i + t(v)(\lambda_1\lambda_2 + \lambda_2\lambda_3 + \lambda_3\lambda_1),$$

where

$$t(v) = -\frac{4\ \mathrm{meas}(e)}{|b_1b_2|^2 + |b_2b_3|^2 + |b_3b_1|^2}\Delta v(G) = -\frac{4\int_e \Delta v dx}{|b_1b_2|^2 + |b_2b_3|^2 + |b_3b_1|^2}.$$

The Carey element space on Ω_h, denoted $\mathcal{C}_h$, is defined by

$$\mathcal{C}_h = \{v:\ v|_e \in \mathcal{C}_e,\ \forall\ e \in \Omega_h\}.$$

1.3.2 Crouzeix–Raviart element

The Crouzeix–Raviart element space[27] on e is denoted $\mathcal{CR}_e$. At a_i, the nodal basis is $1-2\lambda_i$; see Fig.1.3.2. Here, a_1, a_2, and a_3 are the interpolation points. The shape function for e is defined by

$$v = \sum_{i=1}^{3} v(a_i)(1-2\lambda_i).$$

The Crouzeix–Raviart element space on Ω_h, denoted $\mathcal{CR}_h$, is defined by

$$\mathcal{CR}_h = \{v:\ v|_e \in \mathcal{CR}_e,\ \forall\ e \in \Omega_h\}.$$

The Wilson element space[86] and the Wilson-like element space[55] require every element of Ω_h be a convex quadrilateral. And the quartic rectangular element space[53] requires every element of Ω_h be a rectangle. Let $\hat{e} = ([-1,1])^2$ be a reference rectangular element. $\hat{b}_1 = (1,1)$, $\hat{b}_2 = (-1,1)$, $\hat{b}_3 = (-1,-1)$, $\hat{b}_4 = (1,-1)$ are the four vertices of $\hat{e}$. $\hat{a}_1 = (1,0)$, $\hat{a}_2 = (0,1)$, $\hat{a}_3 = (-1,0)$, $\hat{a}_4 = (0,-1)$ are the four edge midpoints of $\hat{e}$. $\hat{a}_5 = \hat{b}_5 = (0,0)$ is the central point of $\hat{e}$. See Fig.1.3.3 and Fig.1.3.4.

1.3.3 Wilson element

The Wilson element space on $\hat{e}$ is denoted $\mathcal{W}_{\hat{e}}$. At $\hat{b}_1$, $\hat{b}_2$, $\hat{b}_3$ and $\hat{b}_4$, its standard nodal bases are

$$\hat{\phi}_1(\xi_1,\xi_2) = \frac{1}{4}(1+\xi_1)(1+\xi_2),$$
$$\hat{\phi}_2(\xi_1,\xi_2) = \frac{1}{4}(1-\xi_1)(1+\xi_2),$$
$$\hat{\phi}_3(\xi_1,\xi_2) = \frac{1}{4}(1-\xi_1)(1-\xi_2),$$
$$\hat{\phi}_4(\xi_1,\xi_2) = \frac{1}{4}(1+\xi_1)(1-\xi_2).$$

At $\hat{b}_5$, it has two bases:

$$\hat{\phi}_{51}(\xi_1,\xi_2) = \frac{1}{2}(\xi_1^2-1), \qquad \hat{\phi}_{52}(\xi_1,\xi_2) = \frac{1}{2}(\xi_2^2-1).$$

Here, $\hat{b}_1, \cdots, \hat{b}_5$ are the interpolation points; see Fig.1.3.3. $\forall\ \hat{v} \in \mathcal{W}_{\hat{e}}$, $\hat{v}$ is uniquely determined by its values at $\hat{b}_1$, $\hat{b}_2$, $\hat{b}_3$, $\hat{b}_4$, and two derivatives of second order at $\hat{b}_5$, i.e.

$$\hat{v}(\xi_1,\xi_2) = \sum_{i=1}^{4} \hat{v}(\hat{b}_i)\hat{\phi}_i(\xi_1,\xi_2) + \frac{\partial^2 \hat{v}}{\partial \xi_1^2}(\hat{b}_5)\hat{\phi}_{51}(\xi_1,\xi_2) + \frac{\partial^2 \hat{v}}{\partial \xi_2^2}(\hat{b}_5)\hat{\phi}_{52}(\xi_1,\xi_2).$$

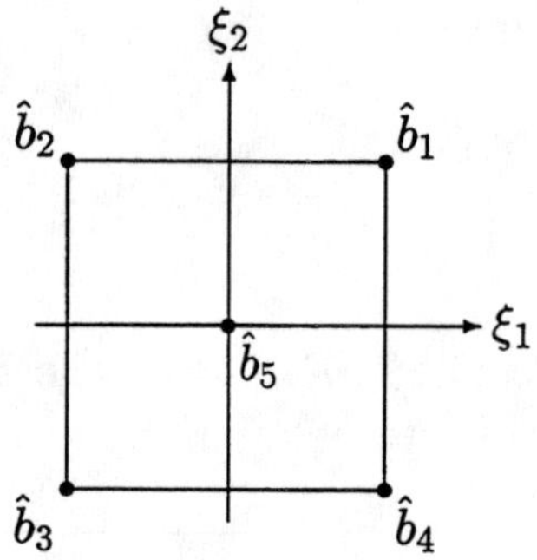

Fig.1.3.3

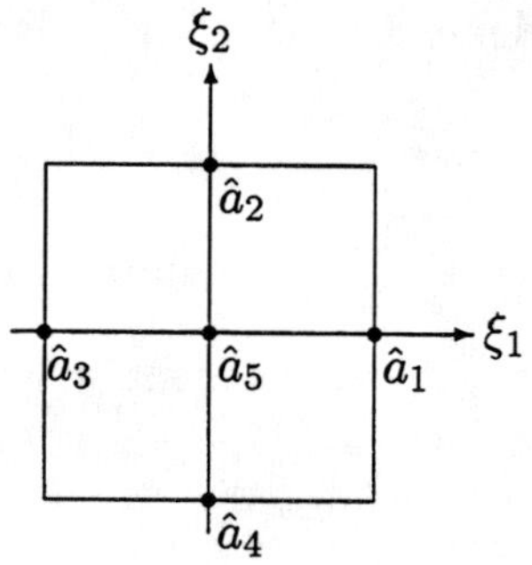

Fig.1.3.4

1.3.4 Wilson–like element

The Wilson–like element space on $\hat{e}$ is denoted $\mathcal{W}l_{\hat{e}}$. At $\hat{b}_1$, $\hat{b}_2$, $\hat{b}_3$ and $\hat{b}_4$, its standard nodal bases are

$$\hat{\psi}_1(\xi_1,\xi_2) = \frac{1}{4}(1+\xi_1)(1+\xi_2),$$

$$\hat{\psi}_2(\xi_1,\xi_2) = \frac{1}{4}(1-\xi_1)(1+\xi_2),$$

$$\hat{\psi}_3(\xi_1,\xi_2) = \frac{1}{4}(1-\xi_1)(1-\xi_2),$$

$$\hat{\psi}_4(\xi_1,\xi_2) = \frac{1}{4}(1+\xi_1)(1-\xi_2).$$

They are the same as these of Wilson element space, but at $\hat{b}_5$, it has two bases of higher order:

$$\hat{\psi}_{51}(\xi_1,\xi_2) = \frac{1}{2}(\xi_1^2-1) - \frac{5}{12}(\xi_1^4-1), \qquad \hat{\psi}_{52}(\xi_1,\xi_2) = \frac{1}{2}(\xi_2^2-1) - \frac{5}{12}(\xi_2^4-1).$$

Here, $\hat{b}_1, \cdots, \hat{b}_5$ are the interpolation points; see Fig.1.3.3. $\forall\, \hat{v} \in \mathcal{W}l_{\hat{e}}$, $\hat{v}$ is uniquely determined by its values at $\hat{b}_1$, $\hat{b}_2$, $\hat{b}_3$, $\hat{b}_4$, and two derivatives of second order at $\hat{b}_5$, i.e.

$$\hat{v}(\xi_1,\xi_2) = \sum_{i=1}^{4} \hat{v}(\hat{b}_i)\hat{\psi}_i(\xi_1,\xi_2) + \frac{\partial^2 \hat{v}}{\partial \xi_1^2}(\hat{b}_5)\hat{\psi}_{51}(\xi_1,\xi_2) + \frac{\partial^2 \hat{v}}{\partial \xi_2^2}(\hat{b}_5)\hat{\psi}_{52}(\xi_1,\xi_2).$$

1.3.5 Quartic rectangular element

The quartic rectangular element space on $\hat{e}$ is denoted $\mathcal{H}_{\hat{e}}$. Let

$$\psi(t) = \frac{1}{2}(5t^4 - 3t^2).$$

At the interpolation points $\hat{a}_1, \hat{a}_2, \cdots, \hat{a}_5$ (see Fig.1.3.4), the standard nodal bases are

$$\hat{\sigma}_1(\xi_1,\xi_2) = \frac{1}{2}\xi_1 + \frac{1}{2}\psi(\xi_1),$$

$$\hat{\sigma}_2(\xi_1,\xi_2) = \frac{1}{2}\xi_2 + \frac{1}{2}\psi(\xi_2),$$

$$\hat{\sigma}_3(\xi_1,\xi_2) = -\frac{1}{2}\xi_1 + \frac{1}{2}\psi(\xi_1),$$

$$\hat{\sigma}_4(\xi_1,\xi_2) = -\frac{1}{2}\xi_2 + \frac{1}{2}\psi(\xi_2),$$

$$\hat{\sigma}_5(\xi_1,\xi_2) = 1 - \psi(\xi_1) - \psi(\xi_2).$$

$\forall\ \hat{v} \in \mathcal{H}_{\hat{e}}$, $\hat{v}$ is uniquely determined by its values at $\hat{a}_1$, $\hat{a}_2$, $\cdots$, $\hat{a}_5$, i.e.

$$\hat{v}(\xi_1,\xi_2) = \sum_{i=1}^{5} \hat{v}(\hat{a}_i)\hat{\sigma}_i(\xi_1,\xi_2).$$

On arbitrary element e, the Wilson element space, the Wilson–like element space and the quartic rectangular element space are denoted by $\mathcal{W}_e$, $\mathcal{W}l_e$ and $\mathcal{H}_e$ respectively. Their bases are induced by the invertible linear or bilinear mapping F_e from $\hat{e}$ to e. For example, if the coordinates of the vertices b_i, $(i = 1,2,3,4)$ of e are (x_{1i}, x_{2i}), then for $\mathcal{W}_e$, the invertible bilinear mapping F_e can be expressed as

$$\begin{cases} x_1 = x_1(\xi_1,\xi_2) = \sum_{i=1}^{4} \hat{\phi}_i(\xi_1,\xi_2)x_{1i} \\ x_2 = x_2(\xi_1,\xi_2) = \sum_{i=1}^{4} \hat{\phi}_i(\xi_1,\xi_2)x_{2i} \end{cases}$$

With F_e, there is a one-to-one relationship between $v \in \mathcal{W}_e$ and $\hat{v} \in \mathcal{W}_{\hat{e}}$, such that

$$\hat{v}(\xi_1,\xi_2) = v(x_1(\xi_1,\xi_2), x_2(\xi_1,\xi_2)),$$

Hence

$$\hat{v} = v \cdot F_e$$

The Wilson element space $\mathcal{W}_e$ is constructed accordingly as

$$\mathcal{W}_e = \{v :\ v \cdot F_e \in \mathcal{W}_{\hat{e}}\}.$$

Associated with Ω_h, the Wilson element space, the Wilson–like element space and the quartic rectangular element space are defined by

$$\mathcal{W}_h = \{v :\ v|_e \in \mathcal{W}_e,\ \forall\ e \in \Omega_h\},$$

$$\mathcal{W}l_h = \{v :\ v|_e \in \mathcal{W}l_e,\ \forall\ e \in \Omega_h\},$$

$$\mathcal{H}_h = \{v :\ v|_e \in \mathcal{H}_e,\ \forall\ e \in \Omega_h\}.$$

It is easy to see that the Wilson element space $\mathcal{W}_h$ has the property that for any $v \in \mathcal{W}_h$, v is continuous at the vertices of each $e \in \Omega_h$. So have the Carey element space $\mathcal{C}_h$ and the Wilson–like element space $\mathcal{W}l_h$. In this book, $\mathcal{W}_h$, $\mathcal{C}_h$ and $\mathcal{W}l_h$ are called *the nonconforming element spaces of the first kind.*

Similarly, we notice that the Crouzeix–Raviart element space $\mathcal{CR}_h$ has the property that for any $v \in \mathcal{CR}_h$, v is continuous at the edge midpoints of each $e \in \Omega_h$. So has the quartic rectangular element space $\mathcal{H}_h$. In this book, $\mathcal{CR}_h$ and $\mathcal{H}_h$ are called *the nonconforming element spaces of the second kind.* These five nonconforming element spaces can be written in the following general form

$$V_h = \Big\{v: \ v|_e \text{ is a polynomial of finite order }, \ v \text{ is continuous at the vertices (edge midpoints) of } e, \quad \forall\, e \in \Omega_h\Big\},$$

$$V_h^0 = \Big\{v \in V_h: \ v(x) = 0, \ \forall \text{ interpolation point } x \in \partial\Omega\Big\}.$$

1.4 Preconditioned conjugate gradient method

The iterative methods, such as the Jacobi, Gauss–Seidel, SOR, Block Jacobi and Block Gauss–Seidel methods, can be viewed as subspace correction methods. The approximate solution is updated by solving problems associated with a set of subspaces of the solution space. These methods can be accelerated by using the conjugate gradient method or the Chebyshev semi–iterative method; we concentrate on the conjugate gradient method.

The conjugate gradient (CG) method is of fundamental importance for the domain decomposition methods. All algorithms studied in this book are variations of preconditioned conjugate gradient method (PCG). Since CG and PCG are well known algorithms, we only give a brief description. For more complete treatment, see [12,39]. We note that in the conjugate gradient method, we only need to be able to apply the matrix A to a given vector; an explicit representation of the matrix is not needed. This is very important for domain decomposition algorithms.

Algorithm 1.4.1.

Set $k = 0$; $x_0 = 0$; $r_0 = b$.
While $|r_k| \geq \epsilon |r_0|$
  $k := k + 1$
  If $k = 1$
    $p_1 = r_0$
  else
    $\beta_k = (r_{k-1}, r_{k-1})/(r_{k-2}, r_{k-2})$
    $p_k = r_{k-1} + \beta_k p_{k-1}$
  end
  $\alpha_k = (r_k, r_k)/(p_k, Ap_k)$
  $x_k = x_k + \alpha_k p_k$

$r_k = r_k - \alpha_k A p_k$

end

The reduction in the energy norm of the error, after n steps of the conjugate gradient iterations, is given by

$$\|x - x_n\| \leq 2\Big(\frac{\sqrt{\kappa}-1}{\sqrt{\kappa}+1}\Big)^n \|x - x_0\|,$$

which can be seen in [12]. Here $\kappa = \kappa(A)$ is the condition number of A given by

$$\kappa = \frac{\lambda_{\max}(A)}{\lambda_{\min}(A)}.$$

The conjugate gradient method is therefore often an effective iterative algorithm to solve symmetric, positive definite systems

$$Ax = b.$$

The algorithm converges to a good approximate solution in relatively few iterations for a well conditioned matrix A. When A is not well conditioned, which is generally the case for discretizations of elliptic problems, we can introduce a preconditioner B and solve the preconditioned linear system

$$B^{-1}Ax = B^{-1}b,$$

by using the conjugate gradient method with the A–inner product. Domain decomposition is an effective way to construct such preconditioners. Other preconditioners are also used in practices, e.g. the diagonal scaling and incomplete Cholesky preconditioners, the multigrid-based preconditioners (see [7,16,17,18,52,54,64,70-73]).

Part I Domain Decomposition Methods for Elliptic Problems of Second Order

Chapter 2

Estimates in Domain Decomposition Analysis

2.1 Purpose

For simplicity of the exposition, we consider the elliptic boundary value problem on a bounded open polygonal domain $\Omega \subset \Re^2$

$$u \in H^1(\Omega): \begin{cases} a(u,v) = (f,v), & \forall\, v \in H_0^1(\Omega) \\ u = g, & \text{on } \partial\Omega \end{cases} \tag{2.1.1}$$

where

$$a(u,v) = \int_\Omega \nabla u \nabla v, \quad (f,v) = \int_\Omega fv, \quad f \in H^{-1}(\Omega), \quad g \in H^{\frac{1}{2}}(\partial\Omega).$$

It is well–known that (2.1.1) has a unique solution $u \in H^1(\Omega)$ (cf.[40,60,68]).

Suppose that $\Omega_h = \{e\}$ is a quasi–uniform mesh of Ω. Let V_h be the nonconforming finite element space on Ω_h of the first or second kind defined in Chapter 1 and π_h the corresponding interpolation operator. V_h has the following general form

$$V_h = \Big\{ v:\ v|_e \text{ is a polynomial of finite order },\ v \text{ is continuous at the}$$

$$\text{vertices (edge midpoints) of } e, \quad \forall\, e \in \Omega_h \Big\},$$

$$V_h^0 = \Big\{ v \in V_h:\ v(x) = 0,\ \forall \text{ interpolation point } x \in \partial\Omega \Big\}.$$

Denote

$$A(w,v) = \sum_{e \in \Omega_h} \int_e \nabla w \nabla v, \quad |v|_{1,\Omega,h} = \sqrt{A(v,v)}.$$

Obviously, $A(\cdot,\cdot), |\cdot|_{1,\Omega,h}$ are the inner product of V_h^0 and its induced norm respectively. The nonconforming finite element discrete problem of (2.1.1) is

$$u_h \in V_h: \begin{cases} A(u_h,v) = (f,v), & \forall\, v \in V_h^0 \\ u_h(x) = g(x), & \forall \text{ interpolation point } x \in \partial\Omega \end{cases} \tag{2.1.2}$$

With the development of parallel computers, domain decomposition methods have recently become an important focus in the field of computational mathematics. By now, all kinds of domain decomposition algorithms have been developed to solve the algebraic system of equations arising from the discretization of (2.1.1) via the conforming finite element methods. It is noted that several fundamental inequalities, especially the so-called extension theorems play key roles in the theoretical analysis of those nonoverlap domain decomposition algorithms (substructuring methods), see, e.g., [12,84]. Therefore, when considering the nonconforming finite element discrete problem (2.1.2), we should establish those inequalities in V_h correspondingly. For this purpose, the conforming interpolation operator I_h is introduced to act as a bridge between V_h and the piecewise linear continuous finite element space where many inequalities have already been constructed[12,26]. Since the regularity of the solution u of (2.1.1) depends on the domain Ω (cf.[40,60,68]), we investigate further error estimations of the nonconforming approximate solution u_h of (2.1.1) under weaker assumption on the regularity. In this way, we eventually establish a series of essential estimates in V_h, some of which are the extension theorems[41,42,50], the Poincaré inequalities and the maximum norm estimate.

The remainder of this chapter is organized as follows: Sect.2.2 gives further error estimations of (2.1.2). Sect.2.3 introduces the conforming interpolation operator I_h and analyses its properties. Sect.2.4 describes and proves some essential estimates in V_h.

2.2 Further error estimations

Theorem 2.2.1. *Let u and u_h be the solutions of (2.1.1) and (2.1.2) respectively. Then*

$$|u - u_h|_{1,\Omega,h} \leq c\Big\{ \inf_{v \in V_h^*} |u - v|_{1,\Omega,h} + \sup_{w \in V_h^0 \setminus \{0\}} \frac{|E(u,w)|}{|w|_{1,\Omega,h}} \Big\}, \tag{2.2.1}$$

where

$$E(u,w) \triangleq A(u,w) - (f,w) = \sum_{e \in \Omega_h} \int_{\partial e} \frac{\partial u}{\partial \mathbf{n}} w, \tag{2.2.2}$$

$$V_h^* = \Big\{ v \in V_h : \ v(x) = u(x), \ \forall \ \textit{interpolation point } x \in \partial\Omega \Big\},$$

$\mathbf{n} = (\nu_1, \nu_2)^T$ represents the unit outward normal vector of e.

Theorem 2.2.1 is in fact the variant of the second Strang lemma[26] in the non-homogeneous boundary value case. Its proof is trivial, so we omit it here.

Let L_e be the linear (bilinear) interpolation operator on e with the vertices of e as its interpolation points. For any measurable set z , we define the mean value operator $M_z : L^2(z) \to \Re$ by

$$\forall \ v \in L^2(z), \quad M_z v \in \Re, \quad M_z v = \frac{1}{\text{meas}(z)} \int_z v.$$

Let $\hat{e}$ be the reference element, which is a square or an isosceles right triangle with $O(1)$ as its area. If there exists an invertible affine mapping

$$x = F_{\hat{e}}(\hat{x}) = B\hat{x} + b : \quad \hat{e} \to e \tag{2.2.3}$$

such that $e = F_{\hat{e}}(\hat{e})$, then we say that e is affine equivalent to $\hat{e}$. Here, $B \in \Re^{2\times 2}$ is nonsingular. For any function v defined on e, let $\hat{v}$ be the corresponding function defined on $\hat{e}$ such that

$$\hat{v}(\hat{x}) = \Big(v \cdot F_{\hat{e}}\Big)(\hat{x}), \quad \forall\, \hat{x} \in \hat{e}.$$

Lemma 2.2.2. *Let $u \in H^{1+\varepsilon}(\Omega)$ be the solution of (2.1.1) ($\varepsilon > 0$). Then*
1) For the first kind nonconforming element space V_h, $\forall\, w \in V_h^0$, we have

$$E(u,w) \le c\Big\{\sum_{e\in\Omega_h}\sum_{i=1}^{2}\int_{\partial e}\Big|\frac{\partial u}{\partial x_i} - M_e\Big(\frac{\partial u}{\partial x_i}\Big)\Big|^2\Big\}^{\frac{1}{2}}\Big(\sum_{e\in\Omega_h}\int_{\partial e}|w - L_e w|^2\Big)^{\frac{1}{2}}. \tag{2.2.4}$$

2) For the second kind nonconforming element space V_h, $\forall\, w \in V_h^0$, we have

$$E(u,w) \le c\Big\{\sum_{e\in\Omega_h}\sum_{i=1}^{2}\int_{\partial e}\Big|\frac{\partial u}{\partial x_i} - M_e\Big(\frac{\partial u}{\partial x_i}\Big)\Big|^2\Big\}^{\frac{1}{2}}\Big(\sum_{e\in\Omega_h}\int_{\partial e}|w - M_e w|^2\Big)^{\frac{1}{2}}. \tag{2.2.5}$$

Proof. Let's prove (2.2.4) first. $\forall\, w \in V_h^0$, $L_e w$ is a linear (bilinear) function on e, and for each $e \in \Omega_h$, we have $L_e w$, which results in a piecewise linear (bilinear) function on Ω_h, denoted $L_e w$ still. Obviously, $L_e w \in H^1(\Omega)$, thus

$$E(u, L_e w) = 0, \; E(u,w) = E(u, w - L_e w).$$

On the other hand, one–by–one analysis shows that

$$\int_{\partial e} \nu_i (w - L_e w) = 0, \quad i = 1,2. \tag{2.2.6}$$

Furthermore, (2.2.4) can be established by the Schwarz inequality.

Now, let's take the Carey membrane element as an example to prove (2.2.6). Let (x_i, y_i) ($i = 1,2,3$) be the vertices of the triangle element e. F_i denotes the opposite edge of (x_i, y_i) whose length is l_i. Let λ_i be the corresponding area coordinate. The unit outward normal vector $\mathbf{n} = (\nu_1, \nu_2)^T$ of e is

$$\nu_1 = \frac{y_{i+2} - y_{i+1}}{l_i}, \quad \nu_2 = -\frac{x_{i+2} - x_{i+1}}{l_i}, \quad \text{on } F_i,\ i = 1,2,3, \tag{2.2.7}$$

where $x_4 = x_1,\ y_4 = y_1,\ x_5 = x_2,\ y_5 = y_2$ for notational convenience. Obviously, in order to see (2.2.6) is right for the Carey element, it suffices to show

$$\int_{\partial e} \nu_i (\lambda_1\lambda_2 + \lambda_2\lambda_3 + \lambda_3\lambda_1) = 0, \quad i = 1,2. \tag{2.2.8}$$

Note that, from (2.2.7), we can obtain

$$\begin{aligned}\sum_{i=1}^{3}\int_{F_i} \nu_1(\lambda_1\lambda_2 + \lambda_2\lambda_3 + \lambda_3\lambda_1) &= \sum_{i=1}^{3}\int_{F_i} \nu_1 \lambda_{i+1}\lambda_{i+2} \\ = \sum_{i=1}^{3} \frac{y_{i+2} - y_{i+1}}{l_i}\int_0^{l_i} \frac{s}{l_i}\frac{l_i - s}{l_i} ds &= \sum_{i=1}^{3} \frac{1}{6}(y_{i+2} - y_{i+1}) = 0.\end{aligned}$$

Thus, (2.2.8) is true for $i = 1$. Similarly, (2.2.8) is true for $i = 2$.

By now, we get (2.2.4).

Next, we prove (2.2.5). For the Crouzeix–Raviart elements, let F be the edge of the triangle element $e \in \Omega_h$. Obviously, $\int_F \nu_i w = 0,\ i = 1, 2,\ \forall\ F \subset \partial\Omega$; if F is the common edge of e_1 and e_2, then it follows from the linearity of w on F and the continuity of w at the midpoint of F that $\int_F \nu_i w_F = 0,\ i = 1, 2$. Here, w_F is the jump of w on F. Therefore, we have

$$\begin{aligned} E(u,w) &= \sum_{e\in\Omega_h} \int_{\partial e} \frac{\partial u}{\partial \mathbf{n}} w = \sum_{e\in\Omega_h} \sum_{F\subset\partial e} \int_F \frac{\partial u}{\partial \mathbf{n}} w = \sum_{e\in\Omega_h} \sum_{F\subset\partial e} \int_F \Big[\frac{\partial u}{\partial \mathbf{n}} - M_F\Big(\frac{\partial u}{\partial \mathbf{n}}\Big)\Big] w \\ &= \sum_{e\in\Omega_h} \sum_{F\subset\partial e} \int_F \Big[\frac{\partial u}{\partial \mathbf{n}} - M_F\Big(\frac{\partial u}{\partial \mathbf{n}}\Big)\Big](w - M_F w) \\ &\le \sum_{e\in\Omega_h} \Big\{\sum_{F\subset\partial e} \int_F \Big|\frac{\partial u}{\partial \mathbf{n}} - M_F\Big(\frac{\partial u}{\partial \mathbf{n}}\Big)\Big|^2\Big\}^{\frac12} \Big(\sum_{F\subset\partial e} \int_F |w - M_F w|^2\Big)^{\frac12} \\ &\le \sum_{e\in\Omega_h} \Big\{\sum_{F\subset\partial e} \sum_{i=1}^{2} \int_F \Big|\frac{\partial u}{\partial x_i} - M_e\Big(\frac{\partial u}{\partial x_i}\Big)\Big|^2\Big\}^{\frac12} \Big(\sum_{F\subset\partial e} \int_F |w - M_e w|^2\Big)^{\frac12}, \end{aligned}$$

which implies that (2.2.5) holds.

For the quartic rectangular elements, let $F_i,\ i = 1, 2, 3, 4$, be the four edges of $e \in \Omega_h$ and a_i their corresponding midpoints. l_i is the length of F_i. a_5 denotes the centroid of e. There exists an invertible affine mapping $x = F_{\hat{e}}(\hat{x}) : \hat{e} \to e$, where $\hat{e}$ is the reference element with $\hat{a}_1 = (1, 0), \hat{a}_2 = (0, 1), \hat{a}_3 = (-1, 0), \hat{a}_4 = (0, -1)$ and $\hat{a}_5 = (0, 0)$, as shown in Fig.2.2.1.

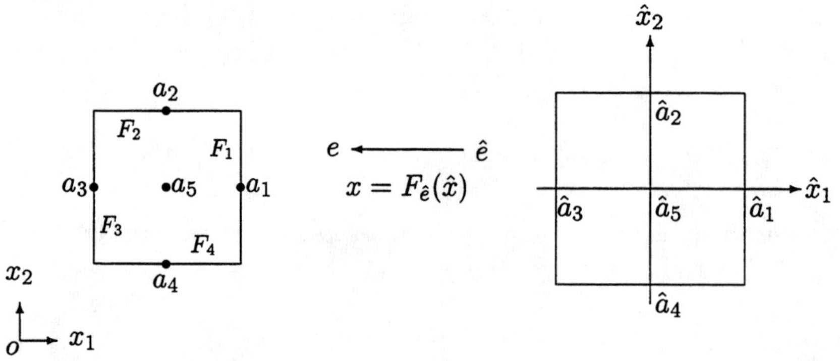

Fig.2.2.1

It is easy to see that

$$E(u,w) = \sum_{e\in\Omega_h} \int_{\partial e} \frac{\partial u}{\partial \mathbf{n}} w = \sum_{e\in\Omega_h} \sum_{i=1}^{4} \int_{F_i} \frac{\partial u}{\partial \mathbf{n}} w = \sum_{e\in\Omega_h} \sum_{i=1}^{4} \int_{F_i} \frac{\partial u}{\partial \mathbf{n}} \big(w - w(a_i)\big). \quad (2.2.9)$$

By the affine mapping $x = F_{\hat{e}}(\hat{x}) : \hat{e} \to e$, we obtain that

$$\int_{F_i} \nu_j\Big(w - w(a_i)\Big) = \frac{l_i}{2}\int_{\hat{F}_i} \nu_j\Big(\hat{w} - \hat{w}(\hat{a}_i)\Big),\ j = 1, 2. \tag{2.2.10}$$

It follows from Fig.2.2.1 and [53] that

$$\int_{\hat{F}_2} \nu_1\Big(\hat{w} - \hat{w}(\hat{a}_2)\Big) = 0, \quad \int_{\hat{F}_4} \nu_1\Big(\hat{w} - \hat{w}(\hat{a}_4)\Big) = 0,$$

$$\hat{w} - \hat{w}(\hat{a}_1) = \frac{\hat{w}(\hat{a}_2) - \hat{w}(\hat{a}_4)}{2}\hat{x}_2 + \frac{\hat{w}(\hat{a}_2) + \hat{w}(\hat{a}_4) - 2\hat{w}(\hat{a}_5)}{2}\phi(\hat{x}_2), \quad \text{on } \hat{F}_1,$$

where

$$\phi(t) = \frac{1}{2}(5t^4 - 3t^2).$$

Since $\int_{-1}^{1} \hat{x}_2 d\hat{x}_2 = 0$ and $\int_{-1}^{1} \phi(\hat{x}_2)d\hat{x}_2 = 0$, we have $\int_{\hat{F}_1} \nu_1\Big(\hat{w} - \hat{w}(\hat{a}_1)\Big) = 0$. In the same manner, it is easy to get $\int_{\hat{F}_3} \nu_1\Big(\hat{w} - \hat{w}(\hat{a}_3)\Big) = 0$. Therefore, by (2.2.10), we obtain that

$$\sum_{i=1}^{4}\int_{F_i} \nu_1\Big(w - w(a_i)\Big) = 0.$$

Similarly, $\sum_{i=1}^{4}\int_{F_i} \nu_2\Big(w - w(a_i)\Big) = 0$ can be established.

Furthermore, it follows from (2.2.9) and the Schwarz inequality that

$$\begin{aligned} E(u, w) &= \sum_{e\in\Omega_h}\sum_{i=1}^{4}\int_{F_i}\Big[\sum_{j=1}^{2}\nu_j\Big(\frac{\partial u}{\partial x_j} - M_e\Big(\frac{\partial u}{\partial x_j}\Big)\Big)\Big]\Big(w - w(a_i)\Big) \\ &\le \sum_{e\in\Omega_h}\Big(\sum_{i=1}^{4}\sum_{j=1}^{2}\int_{F_i}\Big|\frac{\partial u}{\partial x_j} - M_e\Big(\frac{\partial u}{\partial x_j}\Big)\Big|^2\Big)^{\frac{1}{2}}\Big(\sum_{i=1}^{4}\int_{F_i}|w - w(a_i)|^2\Big)^{\frac{1}{2}} \\ &\le \sum_{e\in\Omega_h}\Big(\sum_{i=1}^{4}\sum_{j=1}^{2}\int_{F_i}\Big|\frac{\partial u}{\partial x_j} - M_e\Big(\frac{\partial u}{\partial x_j}\Big)\Big|^2\Big)^{\frac{1}{2}}\Big(\sum_{i=1}^{4}\int_{F_i}|w - M_e w|^2\Big)^{\frac{1}{2}}, \end{aligned}$$

from which (2.2.5) follows. Hence the proof of the lemma is complete. □

Lemma 2.2.3. *For $\varepsilon \in (0, 1)$, we have*

$$\|M_e w\|_{0,e} \le \|w\|_{0,e}, \qquad \forall\ w \in L^2(e) \tag{2.2.11}$$

$$\|w - M_e w\|_{0,e} \le ch_e^{\varepsilon}\|w\|_{\varepsilon,e},\ \forall\ w \in H^{\varepsilon}(e) \tag{2.2.12}$$

Proof. By the Schwarz inequality, it is easy to get (2.2.11). We next prove (2.2.12). Note that $M_e c = c$, $\forall\ c \in \Re$, thus the error estimate of the finite element interpolation[26] yields

$$\|w - M_e w\|_{0,e} \le ch_e|w|_{1,e} \le ch_e\|w\|_{1,e}, \ \ \forall\ w \in H^1(e)$$

It follows from (2.2.11) that

$$\|w - M_e w\|_{0,e} \le c\|w\|_{0,e}, \ \forall\, w \in L^2(e).$$

Therefore, $I - M_e : L^2(e) \to L^2(e)$ and $I - M_e : H^1(e) \to L^2(e)$ are bounded linear operators. Since $H^\varepsilon(e)$ is the interpolation space between $L^2(e)$ and $H^1(e)$, (2.2.12) can be established by the interpolation theorem of Sobolev spaces[1,60]. □

Lemma 2.2.4. *If e is affine equivalent to the reference element* $\hat{e}$ $(1 > \varepsilon \ge \frac{1}{2})$, *then*

$$\int_{\partial e} w^2 \le c\{h_e^{-1}\|w\|_{0,e}^2 + h_e^{2\varepsilon-1}|w|_{\varepsilon,e}^2\}, \ \forall\, w \in H^\varepsilon(e)$$

where $|w|_{\varepsilon,e}^2 \triangleq \int_e\int_e \frac{|w(x) - w(y)|^2}{|x-y|^{2+2\varepsilon}} dxdy.$

Proof. In the reference element $\hat{e}$, the trace theorem[1,60] yields

$$\int_{\partial\hat{e}} (\hat{w})^2 \le c\|\hat{w}\|_{\varepsilon,\hat{e}}^2 \le c\{\|\hat{w}\|_{0,\hat{e}}^2 + |\hat{w}|_{\varepsilon,\hat{e}}^2\}.$$

It follows from the affine equivalence of e and $\hat{e}$ that

$$\|\hat{w}\|_{0,\hat{e}}^2 \le ch_e^{-2}\|w\|_{0,e}^2, \quad |\hat{w}|_{\varepsilon,\hat{e}}^2 \le ch_e^{2\varepsilon-2}|w|_{\varepsilon,e}^2, \quad \int_{\partial\hat{e}} (\hat{w})^2 \ge ch_e^{-1}\int_{\partial e} w^2.$$

With above inequalities, we see that Lemma 2.2.4 holds. □

Lemma 2.2.5. *Let* $u \in H^{1+\varepsilon}(\Omega)$ *be the solution of (2.1.1)* $(1 > \varepsilon \ge \frac{1}{2})$. *We have*

$$E(u, w) \le ch^\varepsilon\|u\|_{H^{1+\varepsilon}(\Omega)}|w|_{1,\Omega,h}, \quad \forall\, w \in V_h^0$$

Proof. (2.2.12) gives

$$\|\frac{\partial u}{\partial x_i} - M_e\Big(\frac{\partial u}{\partial x_i}\Big)\|_{0,e} \le ch_e^\varepsilon\|\frac{\partial u}{\partial x_i}\|_{\varepsilon,e} \le ch_e^\varepsilon\|u\|_{1+\varepsilon,e},$$

$$|\frac{\partial u}{\partial x_i} - M_e\Big(\frac{\partial u}{\partial x_i}\Big)|_{\varepsilon,e} = |\frac{\partial u}{\partial x_i}|_{\varepsilon,e} \le \|u\|_{1+\varepsilon,e}.$$

Combining them with Lemma 2.2.4, we obtain

$$\begin{aligned}
&\sum_{e\in\Omega_h}\sum_{i=1}^{2}\int_{\partial e}|\frac{\partial u}{\partial x_i} - M_e\Big(\frac{\partial u}{\partial x_i}\Big)|^2 \\
&\le c\sum_{e\in\Omega_h}\sum_{i=1}^{2}\Big\{h_e^{-1}\|\frac{\partial u}{\partial x_i} - M_e\Big(\frac{\partial u}{\partial x_i}\Big)\|_{0,e}^2 + h_e^{2\varepsilon-1}|\frac{\partial u}{\partial x_i} - M_e\Big(\frac{\partial u}{\partial x_i}\Big)|_{\varepsilon,e}^2\Big\} \qquad (2.2.13) \\
&\le c\sum_{e\in\Omega_h}\sum_{i=1}^{2}\Big\{h_e^{-1}h_e^{2\varepsilon}\|u\|_{1+\varepsilon,e}^2 + h_e^{2\varepsilon-1}\|u\|_{1+\varepsilon,e}^2\Big\} \le ch^{2\varepsilon-1}\|u\|_{H^{1+\varepsilon}(\Omega)}^2.
\end{aligned}$$

On the other hand, It follows from the interpolation error estimates and the inverse inequalities[26] that

$$\|w - L_e w\|_{0,e} \le ch_e^2|w|_{2,e} \le ch_e|w|_{1,e}, \qquad \|w - M_e w\|_{0,e} \le ch_e|w|_{1,e},$$

$$|w - L_e w|_{1,e} \le ch_e|w|_{2,e} \le c|w|_{1,e}, \qquad |w - M_e w|_{1,e} \le c|w|_{1,e}.$$

Furthermore, Lemma 2.2.4 indicates

$$\begin{aligned}\sum_{e\in\Omega_h}\int_{\partial e}|w-L_ew|^2 &\le c\sum_{e\in\Omega_h}\left\{h_e^{-1}\|w-L_ew\|_{0,e}^2+h_e|w-L_ew|_{1,e}^2\right\}\\ &\le c\sum_{e\in\Omega_h}h_e|w|_{1,e}^2\le ch|w|_{1,\Omega,h}^2.\end{aligned} \tag{2.2.14}$$

Similarly, we have

$$\sum_{e\in\Omega_h}\int_{\partial e}|w-M_ew|^2\le ch|w|_{1,\Omega,h}^2. \tag{2.2.15}$$

Lemma 2.2.5 follows from (2.2.4), (2.2.5), (2.2.13), (2.2.14) and (2.2.15). □

Lemma 2.2.6. *Suppose that* $\hat{\pi}: H^1(\hat{e})\to L^2(\hat{e})$ *is the linear operator on the reference element* $\hat{e}$, *which satisfies that there exists a positive constant* $c_1=c_1(\hat{\pi})$, *such that*

$$\|\hat{\pi}\hat{v}\|_{0,\hat{e}}\le c_1\|\hat{v}\|_{1,\hat{e}},\ \forall\ \hat{v}\in H^1(\hat{e}) \tag{2.2.16}$$

$$\hat{\pi}\hat{c}=\hat{c},\quad\forall\ \hat{c}\in\Re. \tag{2.2.17}$$

Suppose that the element e *is affine equivalent to* $\hat{e}$ *and* π_e *is a linear mapping on* e *defined by*

$$\widehat{\pi_e v}=\hat{\pi}\hat{v},\quad\forall\ v\in H^1(e) \tag{2.2.18}$$

Then there exists a positive constant $c_2=c_2(\hat{\pi},\hat{e})$, *such that*

$$\|v-\pi_e v\|_{0,e}\le c_2h_e|v|_{1,e},\ \forall\ v\in H^1(e)$$

Proof. $\forall\ \hat{c}\in\Re$, (2.2.16) and (2.2.17) yield

$$\|\hat{v}-\hat{\pi}\hat{v}\|_{0,\hat{e}}=\|(\hat{v}+\hat{c})-\hat{\pi}(\hat{v}+\hat{c})\|_{0,\hat{e}}\le(1+c_1)\|\hat{v}+\hat{c}\|_{1,\hat{e}},\ \forall\ \hat{v}\in H^1(\hat{e})$$

Furthermore, it follows from Theorem 3.1.1[26] that

$$\|\hat{v}-\hat{\pi}\hat{v}\|_{0,\hat{e}}\le(1+c_1)\inf_{\hat{c}\in\Re}\|\hat{v}+\hat{c}\|_{1,\hat{e}}\le c_3|\hat{v}|_{1,\hat{e}},\ \forall\ \hat{v}\in H^1(\hat{e})$$

On the other hand, (2.2.18) implies $\hat{v}-\hat{\pi}\hat{v}=\hat{v}-\widehat{\pi_e v}$, hence Theorem 3.1.2[26] gives

$$\|v-\pi_e v\|_{0,e}\le c|\det B|^{\frac{1}{2}}\|\hat{v}-\hat{\pi}\hat{v}\|_{0,\hat{e}},\quad|\hat{v}|_{1,\hat{e}}\le c\|B\||\det B|^{-\frac{1}{2}}|v|_{1,e},$$

where $\det B$ is the determinant of matrix B in the affine equivalence of e and $\hat{e}$. $\|B\|$ represents the Euclidean norm of B in $\Re^2$. It follows from Theorem 3.1.3[26] that $\|B\|\le h_e/\hat{\rho}$, where $\hat{\rho}\stackrel{\triangle}{=}\sup_{B_r\subset\hat{e}}r$. Obviously, $\hat{\rho}=O(1)$.

With the above facts in mind, we end the proof of Lemma 2.2.6. □

Remark 2.2.1. By the imbedding theorem, the Schwarz inequality and the trace theorem, it is easy to see that (2.2.16) is true for finite element interpolation operators.

Let $P(e)$ be the polynomial space on e of finite order. Define the L^2 projection operator $Q_e : L^2(e) \to P(e)$ as follows

$$(Q_e v, w)_{L^2(e)} = (v, w)_{L^2(e)}, \quad \forall\, w \in P(e) \tag{2.2.19}$$

Lemma 2.2.7. *If the element e is affine equivalent to the reference element $\hat{e}$, then we have*

$$|Q_e v|_{1,e} \le c|v|_{1,e}, \quad \forall\, v \in H^1(e) \tag{2.2.20}$$

$$\|v - Q_e v\|_{0,e} \le ch_e|v|_{1,e}, \quad \forall\, v \in H^1(e) \tag{2.2.21}$$

Proof. Analogously to (2.2.19), we can define the L^2 projection operator on $\hat{e}$, which is denoted $Q_{\hat{e}}$. It is easy to see that

$$\widehat{Q_e v} = Q_{\hat{e}}\hat{v},$$

which implies that (2.2.20) is equivalent to

$$|Q_{\hat{e}}\hat{v}|_{1,\hat{e}} \le c|\hat{v}|_{1,\hat{e}}, \quad \forall\, \hat{v} \in H^1(\hat{e}) \tag{2.2.22}$$

It follows from the definition of $Q_{\hat{e}}$ and the Schwarz inequality that

$$\|Q_{\hat{e}}\hat{v}\|_{0,\hat{e}} \le \|\hat{v}\|_{0,\hat{e}}. \tag{2.2.23}$$

Note that the norms of the finite dimensional space $P(\hat{e})$ are equivalent, thus we obtain

$$|Q_{\hat{e}}\hat{v}|_{1,\hat{e}} \le c\|Q_{\hat{e}}\hat{v}\|_{0,\hat{e}} \le c\|\hat{v}\|_{0,\hat{e}} \le c\|\hat{v}\|_{1,\hat{e}},$$

$$|Q_{\hat{e}}\hat{v}|_{1,\hat{e}} = |Q_{\hat{e}}(\hat{v}+\hat{c})|_{1,\hat{e}} \le c\|\hat{v}+\hat{c}\|_{1,\hat{e}}, \ \forall\, \hat{c} \in \Re$$

Furthermore, Theorem 3.1.1[26] yields

$$|Q_{\hat{e}}\hat{v}|_{1,\hat{e}} \le c \inf_{\hat{c}\in\Re} \|\hat{v}+\hat{c}\|_{1,\hat{e}} \le c|\hat{v}|_{1,\hat{e}},$$

which is (2.2.22). Therefore, (2.2.20) holds.

(2.2.23) implies that

$$\|\hat{v} - Q_{\hat{e}}\hat{v}\|_{0,\hat{e}} \le 2\|\hat{v}\|_{0,\hat{e}} \le 2\|\hat{v}\|_{1,\hat{e}},$$

$$\|\hat{v} - Q_{\hat{e}}\hat{v}\|_{0,\hat{e}} = \|(\hat{v}+\hat{c}) - Q_{\hat{e}}(\hat{v}+\hat{c})\|_{0,\hat{e}} \le 2\|(\hat{v}+\hat{c})\|_{1,\hat{e}}, \ \forall\, \hat{c} \in \Re$$

Applying Theorem 3.1.2[26], Theorem 3.1.3[26] and Theorem 3.1.1[26], we have

$$\|v - Q_e v\|_{0,e} \le ch_e\|\hat{v} - Q_{\hat{e}}\hat{v}\|_{0,\hat{e}} \le ch_e \inf_{\hat{c}\in\Re} \|\hat{v}+\hat{c}\|_{1,\hat{e}} \le ch_e|\hat{v}|_{1,\hat{e}} \le ch_e|v|_{1,e}.$$

Hence, (2.2.21) is established. □

Lemma 2.2.8. *Let $\hat{e}, \hat{\pi}, e, \pi_e$ be those in Lemma 2.2.6. $0 \le \varepsilon < 1$. If π_e satisfies*

$$|v - \pi_e v|_{1,e} \le ch_e^k|v|_{k+1,e}, \quad \forall\, v \in H^{k+1}(e) \tag{2.2.24}$$

for some integer $k \geq 1$, then

$$|v - \pi_e v|_{1,e} \leq c h_e^{\varepsilon} |v|_{\varepsilon+1,e}, \quad \forall\ v \in H^{\varepsilon+1}(e) \tag{2.2.25}$$

Proof. It follows from the inverse inequality, Lemma 2.2.6 and Lemma 2.2.7 that

$$\begin{aligned}
|v - \pi_e v|_{1,e} &\leq |v - Q_e v|_{1,e} + |Q_e v - \pi_e v|_{1,e} \\
&\leq c\Big\{|v|_{1,e} + h_e^{-1}\|Q_e v - \pi_e v\|_{0,e}\Big\} \\
&\leq c\Big\{|v|_{1,e} + h_e^{-1}\|Q_e v - v\|_{0,e} + h_e^{-1}\|v - \pi_e v\|_{0,e}\Big\} \\
&\leq c|v|_{1,e}.
\end{aligned}$$

Thus (2.2.25) is true for $\varepsilon = 0$. Furthermore, with (2.2.24) and the interpolation theorem of Sobolev spaces[1,60] , we know that (2.2.25) is true for $\varepsilon > 0$. □

Theorem 2.2.9. *Let $u \in H^{1+\varepsilon}(\Omega)$ be the solution of (2.1.1) ($\varepsilon \geq \frac{1}{2}$). If u_h is the solution of (2.1.2), then*

$$|u - u_h|_{1,\Omega,h} \leq c h^{\varepsilon} \|u\|_{H^{1+\varepsilon}(\Omega)}. \tag{2.2.26}$$

Proof. Note that $\pi_h : H^{1+\varepsilon}(\Omega) \to V_h$ is the finite element interpolation operator, which satisfies

$$(\pi_h w)|_e = \pi_e w, \ \forall\ e \in \Omega_h, \ \forall\ w \in H^{1+\varepsilon}(\Omega).$$

Then

$$\pi_h u \in V_h^*.$$

Since (2.2.24) is the finite element error estimate, by (2.2.25), we get

$$\inf_{v \in V_h^*} |u - v|_{1,\Omega,h} \leq |u - \pi_h u|_{1,\Omega,h} \leq c h^{\varepsilon} |u|_{H^{1+\varepsilon}(\Omega)}.$$

Therefore (2.2.26) follows from Theorem 2.2.1 and Lemma 2.2.5. □

2.3 Conforming interpolation operator

First of all, we construct another mesh $\tilde{\Omega}_h$ of Ω based on Ω_h as follows: for the nonconforming elements of the first kind, if e is a quadrilateral, then e is divided into two triangles by connecting the opposite vertices of e as shown in Fig.2.3.1; for the nonconforming elements of the second kind, e is divided into several triangles by connecting the interpolation points on e as shown in Fig.2.3.2 and Fig.2.3.3. Denote $\tilde{\Omega}_h = \{\tilde{e}\}$ where $\tilde{e} \subset e \in \Omega_h$ is a triangle. Let $S^h(\Omega)$ be the piecewise linear continuous finite element space on $\tilde{\Omega}_h$.

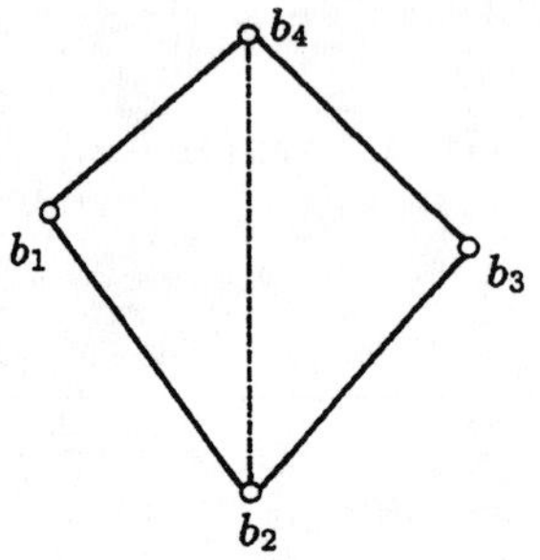

Fig.2.3.1

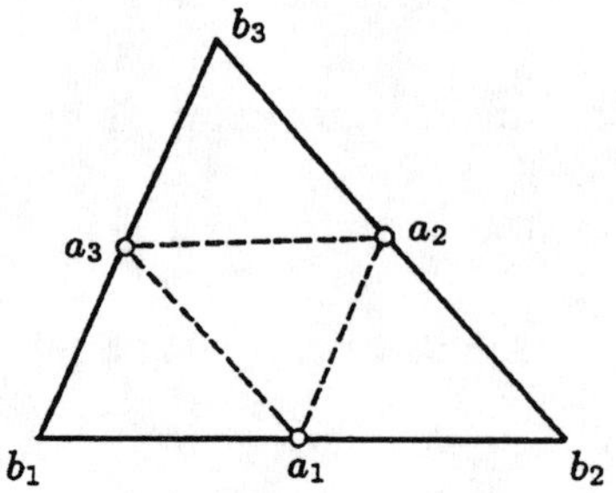

Fig.2.3.2

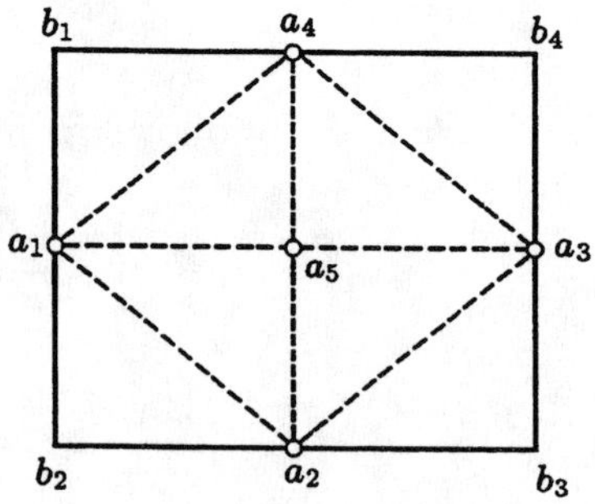

Fig.2.3.3

The conforming interpolation operator $I_h : V_h \to S^h(\Omega)$ is defined as follows

$$\forall\, v \in V_h,\ I_h v \in S^h(\Omega), \quad \text{such that}$$

1) for the nonconforming elements of the first kind,

$$\left(I_h v\right)(b) = v(b), \quad \forall \text{ vertex } b \text{ of } e,\ \forall\, e \in \Omega_h$$

2) for the nonconforming elements of the second kind,

$$\left(I_h v\right)(x) = \begin{cases} v(a), & \forall \text{ interpolation point } x = a \in \overline{\Omega} \\ \frac{1}{2}\left(v(a_1) + v(a_2)\right), & \forall \text{ vertex } x \text{ of } e \in \Omega_h, \\ & \quad x \text{ is not the corner point of } \Omega \\ 0, & \forall \text{ corner point } x \text{ of } \Omega, \\ & \quad a_1, a_2 \in \partial\Omega, v(a_1)\cdot v(a_2) = 0 \\ v(a_1), & \forall \text{ corner point } x \text{ of } \Omega, \\ & \quad a_1, a_2 \in \partial\Omega, v(a_1)\cdot v(a_2) \neq 0 \end{cases}$$

Here, a_1 and a_2 represent the midpoints of any two edges of the elements with x as their common endpoint. Generally, if possible, we select a_1 and a_2 such that a_1, x and a_2 are in a line. But, when $x \in \partial\Omega$ is the vertex of some element e, we select $a_1, a_2 \in \partial\Omega$. Although there might be different way to select a_1 and a_2, we always have

Theorem 2.3.1. *If the conforming interpolation operator I_h on V_h is defined as above, then*

$$\|v - I_h v\|_{L^2(\Omega)} \le ch|v|_{1,\Omega,h}, \quad \forall\, v \in V_h \tag{2.3.1}$$

$$|v - I_h v|_{1,\Omega,h} \le c|v|_{1,\Omega,h}, \quad \forall\, v \in V_h \tag{2.3.2}$$

$$\max_{e\in\Omega_h} \|v - I_h v\|_{L^\infty(e)} \le c|v|_{1,\Omega,h}, \ \forall\, v \in V_h \tag{2.3.3}$$

Proof. $\forall\, v \in V_h$, if (2.3.1) is true, then the inverse inequalities yield

$$\begin{aligned} |v - I_h v|^2_{1,\Omega,h} &= \sum_{e\in\Omega_h} |v - I_h v|^2_{1,e} \\ &\le \sum_{e\in\Omega_h} ch^{-2}|v - I_h v|^2_{0,e} \le ch^{-2}\|v - I_h v\|^2_{L^2(\Omega)} \le c|v|^2_{1,\Omega,h}, \end{aligned}$$

$$\max_{e\in\Omega_h} \|v - I_h v\|_{L^\infty(e)} \le \|v - I_h v\|_{L^\infty(e_0)} \le ch^{-1}\|v - I_h v\|_{0,e_0} \le c|v|_{1,\Omega,h}.$$

Here, $e_0 \in \Omega_h$. Thus, (2.3.1) implies (2.3.2) and (2.3.3). Therefore, it suffices to show (2.3.1).

For the nonconforming elements of the first kind, I_h is the (piecewise) linear interpolation operator on e. It follows from the interpolation error estimate and the inverse inequality that

$$\|v - I_h v\|^2_{L^2(\Omega)} = \sum_{e\in\Omega_h} \|v - I_h v\|^2_{0,e} \le \sum_{e\in\Omega_h} ch^4|v|^2_{2,e} \le \sum_{e\in\Omega_h} ch^2|v|^2_{1,e} \le ch^2|v|^2_{1,\Omega,h},$$

hence, (2.3.1) is established in this case.

For the nonconforming elements of the second kind, construct a function $\tilde{v}$ on Ω, such that $\tilde{v}|_e$ is the unique piecewise linear continuous function determined by $v|_{\tilde{e}}$ and $\tilde{v}|_e$ is linear on $\tilde{e} \subset e$, $\forall\, e \in \Omega_h$. Refer to Fig.2.3.2 and Fig.2.3.3. Obviously, $I_h v = I_h \tilde{v}$, on Ω. The interpolation error estimates and the inverse inequalities yield

$$\|v - \tilde{v}\|^2_{0,e} = \sum_{\tilde{e}\subset e} \|v - \tilde{v}\|^2_{0,\tilde{e}} \le \sum_{\tilde{e}\subset e} ch^4|v|^2_{2,\tilde{e}} = ch^4|v|^2_{2,e} \le ch^2|v|^2_{1,e},$$

$$|v - \tilde{v}|^2_{1,e} = \sum_{\tilde{e}\subset e} |v - \tilde{v}|^2_{1,\tilde{e}} \le \sum_{\tilde{e}\subset e} ch^2|v|^2_{2,\tilde{e}} \le c|v|^2_{1,e},$$

$$\|v - \tilde{v}\|_{L^2(\Omega)} \le ch|v|_{1,\Omega,h}, \tag{2.3.4}$$

$$|v - \tilde{v}|_{1,\Omega,h} \le c|v|_{1,\Omega,h},$$

$$|\tilde{v}|_{1,\Omega,h} \le c|v|_{1,\Omega,h}. \tag{2.3.5}$$

If we can show that

$$\|\tilde{v} - I_h\tilde{v}\|_{L^2(\Omega)} \le ch|\tilde{v}|_{1,\Omega,h}, \tag{2.3.6}$$

then it follows from $I_h v = I_h \tilde{v}$, on Ω, (2.3.4) and (2.3.5) that

$$\begin{aligned} \|v - I_h v\|_{L^2(\Omega)} &\le \|v - \tilde{v}\|_{L^2(\Omega)} + \|\tilde{v} - I_h v\|_{L^2(\Omega)} \\ &= \|v - \tilde{v}\|_{L^2(\Omega)} + \|\tilde{v} - I_h\tilde{v}\|_{L^2(\Omega)} \\ &\le ch|v|_{1,\Omega,h} + ch|\tilde{v}|_{1,\Omega,h} \le ch|v|_{1,\Omega,h}, \end{aligned}$$

which is (2.3.1). Hence, what is left is to prove (2.3.6).

For the quartic rectangular elements, $I_h\tilde{v} = \tilde{v}$, on $\tilde{e}_i$, $i = 5,6,7,8$. Refer to Fig.2.3.4. Therefore, $\|\tilde{v} - I_h\tilde{v}\|^2_{0,e} = \sum_{i=1}^{4} \|\tilde{v} - I_h\tilde{v}\|^2_{0,\tilde{e}_i}$.

On the other hand, on $\overline{e}$, it is easy to see that

$$\tilde{v}(b_1) = v(b_1) = v(a_1) + v(a_4) - v(a_5) = \tilde{v}(a_1) + \tilde{v}(a_4) - \tilde{v}(a_5).$$

If b_1 is not the corner point of Ω, without loss of generality, we assume that

$$\left(I_h\tilde{v}\right)(b_1) = \frac{1}{2}\left(\tilde{v}(a_1) + \tilde{v}(a_6)\right).$$

Then

$$\begin{aligned}
\|\tilde{v} - I_h\tilde{v}\|^2_{0,\tilde{e}_1} &\le ch|\tilde{v}(b_1) - \left(I_h\tilde{v}\right)(b_1)| \\
&\le ch|\tilde{v}(a_1) + \tilde{v}(a_4) - \tilde{v}(a_5) - \tfrac{1}{2}\left(\tilde{v}(a_1) + \tilde{v}(a_6)\right)| \\
&\le ch\left\{|\tilde{v}(a_1) - \tilde{v}(a_5)| + |\tilde{v}(a_4) - \tilde{v}(a_1)| + |\tilde{v}(a_4) - \tilde{v}(a_6)|\right\} \\
&\le ch\left\{|\tilde{v}|_{1,\tilde{e}_5} + |\tilde{v}|_{1,\tilde{e}_9}.\right\}
\end{aligned}$$

If b_1 is the corner point of Ω and $\left(I_h\tilde{v}\right)(b_1) = \tilde{v}(a_1)$, then

$$\|\tilde{v} - I_h\tilde{v}\|_{0,\tilde{e}_1} \le ch|\tilde{v}(b_1) - \left(I_h\tilde{v}\right)(b_1)| \le ch|\tilde{v}(a_4) - \tilde{v}(a_5)| \le ch|\tilde{v}|_{1,\tilde{e}_5}.$$

By now, we obtain that

$$\|\tilde{v} - I_h\tilde{v}\|_{0,e} \le ch\{|\tilde{v}|_{1,e} + |\tilde{v}|_{1,\ddot{e}}\},$$

where $\ddot{e}$ is the union of the elements adjacent to e. Furthermore, summing up with $e \in \Omega_h$, we get (2.3.6).

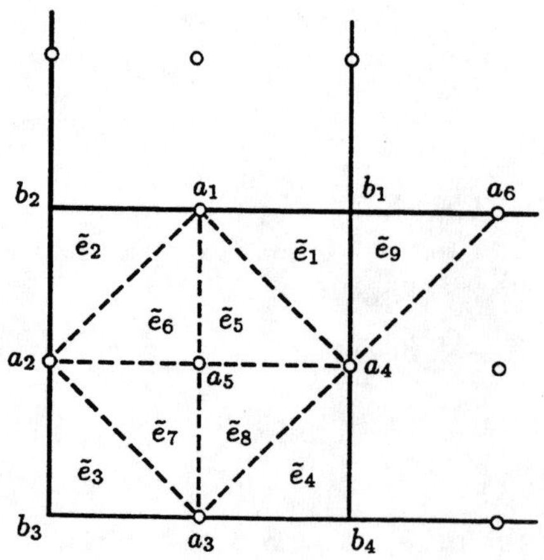

Fig.2.3.4

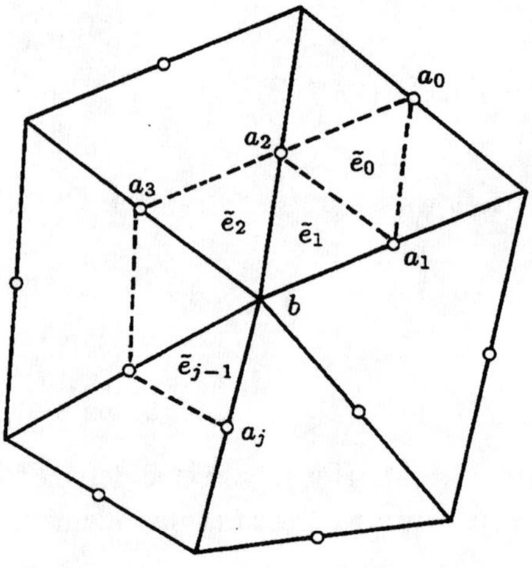

Fig.2.3.5

For the Crouzeix–Raviart elements, $\forall\ e \in \Omega_h,\ \tilde{v} - I_h\tilde{v} = 0$, on $\tilde{e}_0$. Refer to Fig.2.3.5. It suffices to consider $\|\tilde{v} - I_h\tilde{v}\|_{0,\tilde{e}_1}$. Here, $\tilde{e}_0, \tilde{e}_1 \subset e$. In $\bar{e}$, it is easy to see that

$$\tilde{v}(a_1) = \Big(I_h\tilde{v}\Big)(a_1), \quad \tilde{v}(a_2) = \Big(I_h\tilde{v}\Big)(a_2),$$

$$\tilde{v}(b) = v(b) = v(a_1) + v(a_2) - v(a_0) = \tilde{v}(a_1) + \tilde{v}(a_2) - \tilde{v}(a_0).$$

Without loss of generality, assume that

$$\Big(I_h\tilde{v}\Big)(b) = \frac{1}{2}\Big(\tilde{v}(a_i) + \tilde{v}(a_j)\Big),$$

where a_i, a_j $(i \le j)$ are the midpoints of any two edges of the elements with b as their endpoint. It follows from the quasi-uniformness of Ω_h that there exists a positive integer J, independent of h, such that $j \le J$. An elementary calculation yields

$$\begin{aligned}\|\tilde{v} - I_h\tilde{v}\|_{0,\tilde{e}_1} &\le ch|\tilde{v}(b) - \Big(I_h\tilde{v}\Big)(b)| \\ &\le ch|\tilde{v}(a_1) + \tilde{v}(a_2) - \tilde{v}(a_0) - \tfrac{1}{2}\Big(\tilde{v}(a_i) + \tilde{v}(a_j)\Big)| \\ &\le ch\Big\{|\tilde{v}(a_2) - \tilde{v}(a_0)| + |\tilde{v}(a_1) - \tilde{v}(a_i)| + |\tilde{v}(a_1) - \tilde{v}(a_j)|\Big\}.\end{aligned}$$

On the other hand, we have

$$h|\tilde{v}(a_2) - \tilde{v}(a_0)| \le c|\tilde{v}|_{1,\tilde{e}_0},$$

$$h|\tilde{v}(a_1) - \tilde{v}(a_i)| \le h\sum_{k=1}^{i-1}|\tilde{v}(a_k) - \tilde{v}(a_{k+1})| \le c\sum_{k=1}^{i-1}|\tilde{v}|_{1,\tilde{e}_k},$$

$$h|\tilde{v}(a_1) - \tilde{v}(a_j)| \le c\sum_{k=1}^{j-1}|\tilde{v}|_{1,\tilde{e}_k}.$$

Hence

$$\|\tilde{v} - I_h\tilde{v}\|_{0,\tilde{e}_1} \le c\Big\{|\tilde{v}|_{1,\tilde{e}_0} + \sum_{k=1}^{j-1}|\tilde{v}|_{1,\tilde{e}_k}\Big\}.$$

Summing up with $\tilde{e}_1 \subset e,\ e \in \Omega_h$ leads to (2.3.6). By now, the proof of Theorem 2.3.1 is finished. □

2.4 Fundamental inequalities

In this section, $d = \operatorname{diam}\Omega$ represents the diameter of Ω.

Theorem 2.4.1. (Poincaré inequalities in the nonconforming space V_h)

1) $\|v\|^2_{L^2(\Omega)} \le c\Big(d^2|v|^2_{1,\Omega,h} + d^{-2}|\int_\Omega v|^2\Big), \quad \forall\, v \in V_h$

2) $d^{-2}\|v\|^2_{L^2(\Omega)} \le c|v|^2_{1,\Omega,h}, \quad \forall\, v \in V_h^0$

3) *If* $v \in V_h$ *satisfies* $v(x) = 0$, $\forall$ *interpolation point* $x \in \Gamma$, *where* $\Gamma \subset \partial\Omega$ *has at least two interpolation points, then*

$$d^{-2}\|v\|^2_{L^2(\Omega)} \le c|v|^2_{1,\Omega,h}.$$

Proof. $\forall\, v \in V_h$, $I_h v \in S^h(\Omega) \subset H^1(\Omega)$. It follows from the Poincaré inequality in $H^1(\Omega)$ that

$$\begin{aligned}
\|I_h v\|^2_{L^2(\Omega)} &\le c\Big(d^2|I_h v|^2_{H^1(\Omega)} + d^{-2}|\int_\Omega I_h v|^2\Big)\\
&\le c\Big\{d^2|I_h v|^2_{H^1(\Omega)} + d^{-2}|\int_\Omega v|^2 + d^{-2}|\int_\Omega (v - I_h v)|^2\Big\}\\
&\le c\Big\{d^2|v|^2_{1,\Omega,h} + d^2|v - I_h v|^2_{1,\Omega,h} + d^{-2}|\int_\Omega v|^2 + d^2\big(\int_\Omega 1^2\big)\|v - I_h v\|^2_{L^2(\Omega)}\Big\}.
\end{aligned}$$

Furthermore, by the triangle inequality and Theorem 2.3.1, we obtain

$$\begin{aligned}
\|v\|^2_{L^2(\Omega)} &\le 2\Big\{\|v - I_h v\|^2_{L^2(\Omega)} + \|I_h v\|^2_{L^2(\Omega)}\Big\}\\
&\le c\Big\{d^2|v|^2_{1,\Omega,h} + d^{-2}|\int_\Omega v|^2 + h^2|v|^2_{1,\Omega,h}\Big\}.
\end{aligned}$$

Hence, 1) is right.

In the same manner, 2) and 3) can be proved. □

Theorem 2.4.2. (Maximum norm estimate in the nonconforming space V_h)

$$\max_{e\in\Omega_h} \|v\|^2_{L^\infty(e)} \le c\Big\{d^{-2}\|v\|^2_{L^2(\Omega)} + \Big(1 + \ln\frac{d}{h}\Big)|v|^2_{1,\Omega,h}\Big\}, \quad \forall\, v \in V_h$$

Proof. $\forall\, v \in V_h$, $I_h v \in S^h(\Omega) \subset H^1(\Omega)$. It follows from Lemma 3.3[12] and Theorem 2.3.1 that

$$\begin{aligned}
\|I_h v\|^2_{L^\infty(\Omega)} &\le c\Big\{d^{-2}\|I_h v\|^2_{L^2(\Omega)} + \Big(\ln\frac{d}{h}\Big)|I_h v|^2_{H^1(\Omega)}\Big\}\\
&\le c\Big\{d^{-2}\|v\|^2_{L^2(\Omega)} + d^{-2}\|v - I_h v\|^2_{L^2(\Omega)}\\
&\quad + \Big(\ln\frac{d}{h}\Big)|v|^2_{1,\Omega,h} + \Big(\ln\frac{d}{h}\Big)|v - I_h v|^2_{1,\Omega,h}\Big\}\\
&\le c\Big\{d^{-2}\|v\|^2_{L^2(\Omega)} + \Big(\ln\frac{d}{h}\Big)|v|^2_{1,\Omega,h}\Big\},
\end{aligned}$$

$$\begin{aligned}
\max_{e\in\Omega_h} \|v\|^2_{L^\infty(e)} &\le 2\big(\max_{e\in\Omega_h} \|v - I_h v\|^2_{L^\infty(e)} + \|I_h v\|^2_{L^\infty(\Omega)}\big)\\
&\le c\Big\{d^{-2}\|v\|^2_{L^2(\Omega)} + \Big(1 + \ln\frac{d}{h}\Big)|v|^2_{1,\Omega,h}\Big\}.
\end{aligned}$$

By now, the theorem is proved. □

Theorem 2.4.3. (Extension theorem) *Let $\Gamma \subset \partial\Omega$ be an open boundary. $\{\xi_j\}_1^m$ ($m \geq 2$) denotes the set of the interpolation points on Γ (ordered in some way). Give $\lambda = (\lambda_1, \lambda_2, \cdots, \lambda_m)^T \in \Re^m$. If $w_h \in V_h$ satisfies*

$$\begin{cases} A(w_h, v) = 0, & \forall\ v \in V_h^0 \\ w_h(\xi_j) = \lambda_j, & j = 1, 2, \cdots, m \\ w_h(x) = 0, & \forall \text{ interpolation point } x \in \partial\Omega \backslash \Gamma \end{cases} \tag{2.4.1}$$

then

$$c\|\lambda_h\|^2_{H^{\frac{1}{2}}_{00}(\Gamma)} \leq A(w_h, w_h) \leq C\|\lambda_h\|^2_{H^{\frac{1}{2}}_{00}(\Gamma)}, \tag{2.4.2}$$

where λ_h denotes the piecewise linear continuous function on Γ which satisfies

$$\lambda_h(\xi_j) = \lambda_j,\ j = 1, 2, \cdots, m,\ \lambda_h(\nu) = 0,\ \forall \text{ endpoint } \nu \text{ of } \Gamma.$$

Proof. Construct the harmonic function $w \in H^1(\Omega)$, which satisfies

$$\begin{cases} a(w, v) = 0, & \forall\ v \in H_0^1(\Omega) \\ w = \lambda_h, & \text{on } \Gamma \\ w = 0, & \text{on } \partial\Omega \backslash \Gamma \end{cases} \tag{2.4.3}$$

It follows from the priori estimate of the elliptic problems[40,60,68] that even if Ω is concave, there exists $\varepsilon \geq \frac{1}{2}$, such that $w \in H^{\varepsilon+1}(\Omega)$. In addition, we note that w_h is the nonconforming finite element approximation of w.

Theorem 2.2.9 indicates that

$$\begin{aligned} A(w_h, w_h) &\leq 2\{A(w, w) + A(w - w_h, w - w_h)\} \\ &\leq c\{a(w, w) + |w - w_h|^2_{1,\Omega,h}\} \\ &\leq c\{\|w\|^2_{H^1(\Omega)} + h^{2\varepsilon}\|w\|^2_{H^{\varepsilon+1}(\Omega)}\}. \end{aligned}$$

Furthermore, with the priori estimate of the elliptic problems and the fractional order inverse inequality implied by the interpolation theorem of Sobolev spaces, we have

$$\begin{aligned} A(w_h, w_h) &\leq c\{\|w\|^2_{H^{\frac{1}{2}}(\partial\Omega)} + h^{2\varepsilon}\|w\|^2_{H^{\varepsilon+\frac{1}{2}}(\partial\Omega)}\} \\ &\leq c\|w\|^2_{H^{\frac{1}{2}}(\partial\Omega)} \leq c\|w\|^2_{H^{\frac{1}{2}}_{00}(\Gamma)} = c\|\lambda_h\|^2_{H^{\frac{1}{2}}_{00}(\Gamma)}. \end{aligned} \tag{2.4.4}$$

In the above last inequality, we have applied the fact that $\|w\|_{H^{\frac{1}{2}}(\partial\Omega)}$ is equivalent to $\|w\|_{H^{\frac{1}{2}}_{00}(\Gamma)}$ for $w = 0$ on $\partial\Omega \backslash \Gamma$ (cf.[12]).

On the other hand, it follows from the construction of I_h that

$$I_h w_h \in H^1(\Omega),\ I_h w_h = 0, \text{ on } \partial\Omega \backslash \Gamma, \quad I_h w_h = \lambda_h, \text{ on } \Gamma.$$

With Theorem 2.3.1, Lemma 1.2.2 and the trace theorem, we obtain

$$\begin{aligned}A(w_h, w_h) &\geq c|w_h|^2_{1,\Omega,h} \geq c|I_h w_h|^2_{H^1(\Omega)} \geq c\|I_h w_h\|^2_{H^1(\Omega)} \\ &\geq c\|I_h w_h\|^2_{H^{\frac{1}{2}}(\partial\Omega)} \geq c\|I_h w_h\|^2_{H^{\frac{1}{2}}_{00}(\Gamma)} = c\|\lambda_h\|^2_{H^{\frac{1}{2}}_{00}(\Gamma)},\end{aligned}$$

Therefore, (2.4.2) follows from (2.4.4). □

Lemma 2.4.4. *Suppose that $d \gg h$ and there exists a positive constant β, such that $\sup\limits_{B_r \subset \Omega} r \geq \beta d$. Let ν_1 be a corner point of Ω. v is the piecewise linear continuous function on $\partial\Omega$ which satisfies*

$$v(x) = 0, \ \forall \text{ interpolation point } x \in \partial\Omega,$$

$$v(\nu) = 0, \ \forall \text{ corner point } \nu \text{ of } \Omega, \ \nu \neq \nu_1.$$

Then, there exists a positive constant c, independent of d, such that

$$\|v\|_{H^{\frac{1}{2}}(\partial\Omega)} \leq c|v(\nu_1)|. \tag{2.4.5}$$

Proof. Without loss of generality, we assume that Ω is a triangle. Let $\Gamma_1, \Gamma_2, \Gamma_3$ be the three edges of Ω with ν_1, ν_2, ν_3 as their opposite corner points. The definition of the Sobolev space $H^{\frac{1}{2}}(\partial\Omega)$ is (cf.[60])

$$\begin{aligned}\|v\|^2_{H^{\frac{1}{2}}(\partial\Omega)} &\triangleq \frac{1}{\text{meas}(\partial\Omega)}\int_{\partial\Omega} v^2 ds + \int_{\partial\Omega}\int_{\partial\Omega} \frac{|v(x) - v(y)|^2}{|x-y|^2} ds(x)ds(y) \\ &= \frac{1}{\text{meas}(\partial\Omega)}\int_{\partial\Omega} v^2 ds + \sum_{i,j=1}^{3}\int_{\Gamma_i}\int_{\Gamma_j} \frac{|v(x) - v(y)|^2}{|x-y|^2} ds(x)ds(y).\end{aligned}$$

Since $d \gg h$ and $\sup\limits_{B_r \subset \Omega} r \geq \beta d$, a simple calculus calculation yields

$$\frac{1}{\text{meas}(\partial\Omega)}\int_{\partial\Omega} v^2 ds \leq c|v(\nu_1)|^2,$$

$$\begin{aligned}&\int_{\Gamma_1}\int_{\Gamma_1} = 0, && \int_{\Gamma_1}\int_{\Gamma_2} = \int_{\Gamma_2}\int_{\Gamma_1} \leq c|v(\nu_1)|^2, \\ &\int_{\Gamma_2}\int_{\Gamma_2} \leq 2|v(\nu_1)|^2, && \int_{\Gamma_2}\int_{\Gamma_3} = \int_{\Gamma_3}\int_{\Gamma_2} < 2|v(\nu_1)|^2, \\ &\int_{\Gamma_3}\int_{\Gamma_3} \leq 2|v(\nu_1)|^2, && \int_{\Gamma_3}\int_{\Gamma_1} = \int_{\Gamma_1}\int_{\Gamma_3} \leq c|v(\nu_1)|^2.\end{aligned}$$

Here, we omit the integrands for conciseness and c is independent of d, h.

By now, (2.4.5) is established. □

Lemma 2.4.5. *Let Γ_h be the quasi-uniform mesh of the interval $\Gamma = [0, L]$. Suppose that $v(x)$ is the piecewise linear continuous function on Γ_h and $v(0) = 0$. Then, there exists a constant c, independent of L, such that*

$$\int_{\Gamma} \frac{(v(x))^2}{x} dx \leq \left(1 + \ln\frac{L}{h}\right)\|v\|^2_{L^\infty(\Gamma)}. \tag{2.4.6}$$

The proof of Lemma 2.4.5 is trivial, so we omit it here.

Theorem 2.4.6. (Extension theorem) *Suppose that $d \gg h$ and there exists a positive constant β, such that $\sup_{B_r \subset \Omega} r \geq \beta d$. Let $\{\nu_k\}_1^J$ be the set of the corner points of Ω. $\Gamma_{ij} \subset \partial\Omega$ denotes the edge of Ω with ν_i and ν_j as its endpoints. Suppose that V_h is the nonconforming finite element space of the second kind. Then*

$$\|v^{ij}\|^2_{H^{\frac{1}{2}}_{00}(\Gamma_{ij})} \leq c\Big(1+\ln\frac{d}{h}\Big)\Big\{\Big(1+\ln\frac{d}{h}\Big)|v|^2_{1,\Omega,h} + (1+d^{-2})\|v\|^2_{L^2(\Omega)}\Big\}, \ \forall\ v \in V_h$$

where v^{ij} is the piecewise linear continuous function on Γ_{ij} which satisfies

$$v^{ij}(x) = v(x),\ \forall \textit{ interpolation point } x \in \Gamma_{ij},\ v^{ij}(\nu_i) = v^{ij}(\nu_j) = 0,$$

and the positive constant c is independent of d.

Proof. Let v^B be the piecewise linear continuous function on $\partial\Omega$, such that

$$v^B = v^{ij}, \text{on } \Gamma_{ij},\ \forall\ \Gamma_{ij} \subset \partial\Omega$$

Let v_k $(k = 1, 2, \cdots, J)$ be the piecewise linear continuous function on $\partial\Omega$, such that

$$v_k(\nu_k) = \Big(I_h v\Big)(\nu_k),\ v_k(\nu_j) = 0,\ j = 1, 2, \cdots, k-1, k+1, \cdots, J,$$

$$v_k(x) = 0,\ \forall \text{ interpolatoin point } x \in \partial\Omega.$$

It is easy to see that

$$I_h v = v^B + \sum_{k=1}^{J} v_k, \text{on } \partial\Omega.$$

It follows from the definition of the Sobolev space $H^{\frac{1}{2}}_{00}(\Gamma_{ij})$ that[12,60]

$$\|v^{ij}\|^2_{H^{\frac{1}{2}}_{00}(\Gamma_{ij})} \leq c\Big\{\|v^B\|^2_{H^{\frac{1}{2}}(\partial\Omega)} + \int_{\Gamma_{ij}} \Big(\frac{|v^{ij}(x)|^2}{|x-\nu_i|} + \frac{|v^{ij}(x)|^2}{|x-\nu_j|}\Big) ds(x)\Big\}. \tag{2.4.7}$$

With the trace theorem, Lemma 2.4.4, Lemma 3.3[12] and Theorem 2.3.1, we have

$$\begin{aligned}
\|v^B\|^2_{H^{\frac{1}{2}}(\partial\Omega)} &= \|I_h v - \sum_{k=1}^{J} v_k\|^2_{H^{\frac{1}{2}}(\partial\Omega)} \\
&\leq c\{\|I_h v\|^2_{H^{\frac{1}{2}}(\partial\Omega)} + \sum_{k=1}^{J} \|v_k\|^2_{H^{\frac{1}{2}}(\partial\Omega)}\} \\
&\leq c\{\|I_h v\|^2_{H^1(\Omega)} + \sum_{k=1}^{J} |\Big(I_h v\Big)(\nu_k)|^2\} \\
&\leq c\{\|I_h v\|^2_{H^1(\Omega)} + \|I_h v\|^2_{L^\infty(\Omega)}\} \\
&\leq \Big\{\Big(1+\ln\frac{d}{h}\Big)|I_h v|^2_{H^1(\Omega)} + (1+d^{-2})\|I_h v\|^2_{L^2(\Omega)}\Big\}, \\
&\leq \Big\{\Big(1+\ln\frac{d}{h}\Big)|v|^2_{1,\Omega,h} + (1+d^{-2})\|v\|^2_{L^2(\Omega)}\Big\}.
\end{aligned} \tag{2.4.8}$$

It follows from Lemma 2.4.5, Lemma 3.3[12] and Theorem 2.3.1 that

$$\begin{aligned}\int_{\Gamma_{ij}}\Big(\frac{|v^{ij}(x)|^2}{|x-\nu_i|}+\frac{|v^{ij}(x)|^2}{|x-\nu_j|}\Big)ds(x) &\le c\Big(1+\ln\frac{d}{h}\Big)\|v^{ij}\|^2_{L^\infty(\Gamma_{ij})}\\ &\le c\Big(1+\ln\frac{d}{h}\Big)\|I_hv\|^2_{L^\infty(\Omega)}\\ &\le c\Big(1+\ln\frac{d}{h}\Big)\Big\{\Big(\ln\frac{d}{h}\Big)|I_hv|^2_{H^1(\Omega)}+d^{-2}\|I_hv\|^2_{L^2(\Omega)}\Big\}\\ &\le c\Big(1+\ln\frac{d}{h}\Big)\Big\{\Big(\ln\frac{d}{h}\Big)|v|^2_{1,\Omega,h}+d^{-2}\|v\|^2_{L^2(\Omega)}\Big\}.\end{aligned}\tag{2.4.9}$$

(2.4.7), (2.4.8) and (2.4.9) indicate that Theorem 2.4.6 holds. □

Remark 2.4.1. Although we deal with the Dirichlet form $a(u,v)=\int_\Omega \nabla u\nabla v$ in this chapter, all the above conclusions are true for the general form

$$a(u,v)=\int_\Omega\Big[\sum_{i,j=1}^{2}a_{ij}(x)\frac{\partial u}{\partial x_i}\frac{\partial v}{\partial x_j}+a_0(x)uv\Big],$$

where $a_{ij}(x), a_0(x)$ are bounded, piecewise smooth on Ω, $a_0(x)\ge 0$ and (a_{ij}) is a symmetric, uniformly positive definite matrix on Ω.

Remark 2.4.2. Besides [21,55,86], the nonconforming finite element space of the first kind may include other elements which are continuous at the vertices of every $e\in\Omega_h$, if (2.2.6) is true.

Remark 2.4.3. Besides [27,53], the nonconforming finite element space of the second kind may include other elements which are continuous at the midpoints, even the Gaussian quadrature points, of the edges of every $e\in\Omega_h$, if (2.2.5) and (2.3.6) are established.

Chapter 3

Substructuring algorithms with two–subdomains

3.1 Introduction

Although the domain decomposition algorithms with two nonoverlap subdomains (substructures) don't virtually and completely have the advantages of domain decomposition, the study of them is the first step to establish more efficient and complicated algorithms. In the conforming finite element discrete cases, different algorithms are presented for second order elliptic problems[9,11,65,89], and their rate of convergence are bounded independently of the size of the problem, which is ensured by the so–called extension theorem[84]. It remains a problem, which the present chapter will solve, to design and analyse the domain decomposition algorithms in the nonconforming discrete cases. We consider the nonconforming finite elements of the first and second kind defined in Chapter 1. We note that the extension theorems for those nonconforming elements have already been established in Chapter 1, which play a key role in the analysis of domain decomposition methods. Based on the extension theorems, a unifying and systematic way to design and analyse the domain decomposition methods with two substructures are developed for the second order self–adjoint elliptic problems discretized by the finite elements, conforming or nonconforming. In addition, a brief survey of all existing algorithms, some of which are discussed in detail, is presented in this chapter.

An outline of the remainder is as follows: In Sect.3.2, the Steklov–Poincaré operator and the extension theorem in the continuous case is stated in brief, while their counterparts in the finite element discrete case is presented in Sect.3.3. The convergence analysis and matrix analysis are given in Sect.3.4 and Sect.3.5. Finally, in Sect.3.6, we describe the results of numerical calculations which show that the theoretical estimates are fully realized in practice.

3.2 Extension theorem in the continuous cases

Let $\Omega \subset \Re^2$ be an open bounded domain. Consider the second order self–adjoint elliptic problem in the weak formulation

$$u \in H_0^1(\Omega): \ a(u,v) = (f,v), \ \forall\, v \in H_0^1(\Omega) \tag{3.2.1}$$

where, $(f,v) = \int_\Omega fv$, $a(u,v) = \int_\Omega \Big[\sum_{i,j=1}^{2} a_{ij}(x)\frac{\partial u}{\partial x_i}\frac{\partial v}{\partial x_j} + a_0(x)uv\Big]$, $f \in H^{-1}(\Omega)$, $a_0(x) \geq 0$. $(a_{ij}(x))$ is a symmetric, uniformly positive definite matrix. $a_{ij}(x), a_0(x)$

are bounded and piecewise smooth on Ω. Let Ω be divided by an open curve Γ into two open subdomains Ω_1 and Ω_2, such that

$$\Omega = \Omega_1 \cup \Omega_2 \cup \Gamma, \quad \Omega_1 \cap \Omega_2 = \emptyset.$$

For $k = 1, 2$, denote

$$(f, v)_k = \int_{\Omega_k} fv, \quad a_k(u, v) = \int_{\Omega_k} \Big[\sum_{i,j=1}^{2} a_{ij} \frac{\partial u}{\partial x_i}\frac{\partial v}{\partial x_j} + a_0 uv\Big].$$

Once the restriction to Γ of the solution u of (3.2.1), also known as the trace, $\lambda = r_0 u = u|_\Gamma \in H_{00}^{\frac{1}{2}}(\Gamma)$ is obtained,

$$u_k = u|_{\Omega_k} \in H_\Gamma^1(\Omega_k) = \{w \in H^1(\Omega_k) : \; w = 0, \text{ on } \partial\Omega_k \backslash \Gamma\},$$

satisfies independently

$$\begin{cases} a_k(u_k, v) = (f, v)_k, \; \forall \, v \in H_0^1(\Omega_k) \\ u_k = \lambda, \qquad\qquad \textit{on } \Gamma \end{cases} \tag{3.2.2}$$

But λ is unknown. It follows from the Green formula that λ should satisfy the following interface condition

$$\frac{\partial u_1(\lambda)}{\partial \mathbf{n}_1} = -\frac{\partial u_2(\lambda)}{\partial \mathbf{n}_2}, \text{ on } \Gamma. \tag{3.2.3}$$

Here, $\dfrac{\partial}{\partial \mathbf{n}_k} = \sum\limits_{i,j=1}^{2} a_{ij}\nu_i^k \dfrac{\partial}{\partial x_j}$, $(\nu_1^k, \nu_2^k)^T$ is the unit outward normal vector. It is easy to see that the solution u_k of (3.2.2) can be expressed as

$$u_k = E_k\lambda + T_k f, \tag{3.2.4}$$

where $T_k f$ is the solution of

$$T_k f \in H_0^1(\Omega_k) : \; a_k(T_k f, v) = (f, v)_k, \; \forall \, v \in H_0^1(\Omega_k) \tag{3.2.5}$$

and $E_k\lambda \in H_\Gamma^1(\Omega_k)$ is the harmonic extension function on Ω_k determined by λ, which satisfies

$$\begin{cases} a_k(E_k\lambda, v) = 0, \quad \forall \, v \in H_0^1(\Omega_k) \\ E_k\lambda = \lambda, \qquad\quad \text{on } \Gamma \end{cases} \tag{3.2.6}$$

Correspondingly, $E_k : H_{00}^{\frac{1}{2}}(\Gamma) \to H_\Gamma^1(\Omega_k)$ is called *the harmonic extension operator.* The substitution of (3.2.4) into (3.2.3) yields

$$\frac{\partial}{\partial \mathbf{n}_1}(E_1 - E_2)\lambda = \frac{\partial}{\partial \mathbf{n}_1}(T_2 - T_1)f, \text{ on } \Gamma. \tag{3.2.7}$$

For the right hand side of (3.2.7), $\mu = \frac{\partial}{\partial \mathbf{n}_1}(T_2 - T_1)f$, is independent of λ, and can be obtained by solving (3.2.5) previously and independently, (3.2.7) is equivalent to the operator equation

$$S\lambda = \mu, \tag{3.2.8}$$

where $S = S_1 + S_2$, $S_k \cdot = \frac{\partial}{\partial \mathbf{n}_k}(E_k \cdot)$. S is called the Steklov–Poincaré operator[3]. We know that

1) $S : H_{00}^{\frac{1}{2}}(\Gamma) \to H^{-\frac{1}{2}}(\Gamma)$ and $S_k : H_{00}^{\frac{1}{2}}(\Gamma) \to H^{-\frac{1}{2}}(\Gamma)$ are symmetric, positive definite operators;

2) The inverse operator of S_k, denoted $S_k^{-1} : H^{-\frac{1}{2}}(\Gamma) \to H_{00}^{\frac{1}{2}}(\Gamma)$, exists. For any $\mu \in H^{-\frac{1}{2}}(\Gamma)$, $\lambda = S_k^{-1}\mu$ can be obtained by first solving the Neumann–Dirichlet mixed problem on Ω_k:

$$u_k \in H_\Gamma^1(\Omega_k) : \ a_k(u_k, v) = \int_\Gamma \mu v, \ \forall\, v \in H_\Gamma^1(\Omega_k)$$

and then setting $\lambda = u|_\Gamma$.

With the trace theorem and a well–known priori inequality[60,68], we obtain

Theorem 3.2.1. (The extension theorem in the continuous cases) *There exist two constants $\tilde{\sigma}, \tilde{\tau}$, such that*

$$\tilde{\sigma} a_1(E_1\lambda, E_1\lambda) \le a_2(E_2\lambda, E_2\lambda) \le \tilde{\tau} a_1(E_1\lambda, E_1\lambda), \ \forall\, \lambda \in H_{00}^{\frac{1}{2}}(\Gamma) \tag{3.2.9}$$

3.3 Extension theorem in the discrete case

In what follows, we assume that Ω is a polygonal domain. Let $\Omega_h = \{e\}$ be a quasi–uniform mesh of Ω, such that $e \cap \Gamma = \emptyset$, $\forall\, e \in \Omega_h$, where e, a triangle or a quadrilateral, represents the typical element in Ω_h. Let V_h be the finite element space associated with Ω_h. Of course, V_h can be the space of the conforming elements. Also, V_h can be the space of the first or second kind nonconforming elements introduced in Chapter 1. For simplicity of the exposition, an interpolation point x of V_h is briefly called a node, and we assume that

$$v(x) = 0, \ \forall \text{ node } x \in \partial\Omega, \ \forall\, v \in V_h$$

All the nodes on Γ are labelled $\{\xi_j\}_{j=1}^m$, which are ordered in some way. For $k = 1, 2$, we introduce the following notations

$$A(u, v) = \sum_{e \subset \Omega} \int_e \Big[\sum_{i,j=1}^{2} a_{ij} \frac{\partial u}{\partial x_i}\frac{\partial v}{\partial x_j} + a_0 uv\Big],$$

$$A_k(u, v) = \sum_{e \subset \Omega_k} \int_e \Big[\sum_{i,j=1}^{2} a_{ij} \frac{\partial u}{\partial x_i}\frac{\partial v}{\partial x_j} + a_0 uv\Big],$$

$$V_h^k = \{v \in V_h : \ v(x) = 0, \ \forall \text{ node } x \in \Omega \backslash \overline{\Omega}_k\},$$
$$V_h^{k,0} = \{v \in V_h : \ v(x) = 0, \ \forall \text{ node } x \in \Omega \backslash \Omega_k\}.$$

The finite element discrete problem corresponding to (3.2.1) is

$$u_h \in V_h : \ A(u_h, v) = (f, v), \ \forall \ v \in V_h \tag{3.3.1}$$

Let $r_0^h : V_h \to \Re^m$ be the discrete trace operator, such that

$$\forall \ v \in V_h, \ r_0^h v \in \Re^m : \ \left(r_0^h v\right)(j) = v(\xi_j), \ j = 1, 2, \cdots, m,$$

where $\left(r_0^h v\right)(j)$ represents the jth element of $r_0^h v$. We define *the discrete harmonic extension operator* $E_k^h : \Re^m \to V_h^k$ as follows

$$\forall \ \lambda \in \Re^m, \ E_k^h \lambda \in V_h^k : \begin{cases} A_k(E_k^h \lambda, v) = 0, & \forall \ v \in V_h^{k,0} \\ r_0^h E_k^h \lambda = \lambda \end{cases}$$

Theorem 2.4.3 and the extension theorem for conforming elements in [84] can be expressed in the following general form:

Theorem 3.3.1. (The extension theorem in the discrete case) *There exist constants σ, τ, c and C, independent of the quasi–uniform mesh parameter h, such that*

$$\sigma = \sup_{\lambda \in \Re^m \backslash \{0\}} \frac{A_1(E_1^h \lambda, E_1^h \lambda)}{A_2(E_2^h \lambda, E_2^h \lambda)}, \quad \tau = \sup_{\lambda \in \Re^m \backslash \{0\}} \frac{A_2(E_2^h \lambda, E_2^h \lambda)}{A_1(E_1^h \lambda, E_1^h \lambda)}, \tag{3.3.2}$$

$$c \|\lambda_h\|^2_{H^{\frac{1}{2}}_{00}(\Gamma)} \le A_k(E_k^h \lambda, E_k^h \lambda) \le C \|\lambda_h\|^2_{H^{\frac{1}{2}}_{00}(\Gamma)}, \ k = 1, 2, \tag{3.3.3}$$

where λ_h is the piecewise linear continuous function on Γ interpolated by $\{\xi_j\}_{j=1}^m$ with λ, such that

$$\lambda_h(\xi_j) = \lambda_j, \ j = 1, 2, \cdots, m, \quad \lambda_h(\nu) = 0, \ \forall \textit{ endpoint } \nu \text{ of } \Gamma.$$

Furthermore, for any $\lambda \in \Re^m$, define $T_k^h : \Re^m \to V_h^k$ as follows

$$\begin{cases} T_2^h \lambda \in V_h^2 \\ A_2(T_2^h \lambda, v) = -A_1(E_1^h \lambda, E_1^h r_0^h v), \forall \ v \in V_h^2 \end{cases} \tag{3.3.4}$$

$$\begin{cases} T_1^h \lambda \in V_h^1 \\ A_1(T_1^h \lambda, v) = -A_2(E_2^h \lambda, E_2^h r_0^h v), \forall \ v \in V_h^1 \end{cases} \tag{3.3.5}$$

Corrollary 3.3.2. *Let Ω_h be quasi–uniform, σ and τ the same as those in Theorem 3.3.1. Then we have*

$$\frac{1}{\tau} A_1(E_1^h \lambda, E_1^h \lambda) \le A_2(T_2^h \lambda, T_2^h \lambda) \le \sigma A_1(E_1^h \lambda, E_1^h \lambda), \ \forall \ \lambda \in \Re^m \tag{3.3.6}$$

$$\frac{1}{\sigma}A_2(E_2^h\lambda, E_2^h\lambda) \le A_1(T_1^h\lambda, T_1^h\lambda) \le \tau A_2(E_2^h\lambda, E_2^h\lambda),\ \forall\ \lambda \in \Re^m \tag{3.3.7}$$

Proof. It follows from the substitution of v in (3.3.4) with $T_2^h\lambda$, the Schwarz inequality and (3.3.2) that

$$\begin{aligned} A_2(T_2^h\lambda, T_2^h\lambda) &= -A_1(E_1^h\lambda, E_1^h r_0^h T_2^h\lambda) \\ &\le \Big(A_1(E_1^h\lambda, E_1^h\lambda)\Big)^{\frac{1}{2}} \Big(A_1(E_1^h r_0^h T_2^h\lambda, E_1^h r_0^h T_2^h\lambda)\Big)^{\frac{1}{2}} \\ &\le \Big(A_1(E_1^h\lambda, E_1^h\lambda)\Big)^{\frac{1}{2}} \Big(\sigma A_2(E_2^h r_0^h T_2^h\lambda, E_2^h r_0^h T_2^h\lambda)\Big)^{\frac{1}{2}} \\ &= \Big(A_1(E_1^h\lambda, E_1^h\lambda)\Big)^{\frac{1}{2}} \Big(\sigma A_2(T_2^h\lambda, T_2^h\lambda)\Big)^{\frac{1}{2}}, \end{aligned}$$

which implies that the right hand side of (3.3.6) is true.

On the other hand, taking $v = E_2^h\lambda$ in (3.3.4) and using (3.3.2), we get

$$\begin{aligned} A_1(E_1^h\lambda, E_1^h\lambda) = -A_2(T_2^h\lambda, E_2^h\lambda) &\le \Big(A_2(T_2^h\lambda, T_2^h\lambda)\Big)^{\frac{1}{2}} \Big(A_2(E_2^h\lambda, E_2^h\lambda)\Big)^{\frac{1}{2}} \\ &\le \Big(A_2(T_2^h\lambda, T_2^h\lambda)\Big)^{\frac{1}{2}} \Big(\tau A_1(E_1^h\lambda, E_1^h\lambda)\Big)^{\frac{1}{2}}. \end{aligned}$$

Therefore, the left hand side of (3.3.6) is proved.

Similarly, (3.3.7) can be established. □

For $k = 1, 2$, for any $\lambda \in \Re^m$, let $\rho_k : \Re^m \to V_h^k$ be the linear operator, s.t.

$$\rho_k\lambda \in V_h^k,\ \Big(\rho_k\lambda\Big)(x) = \begin{cases} 0, & \forall \text{ node } x \in \Omega_k \\ \lambda_j, & x = \xi_j,\ j = 1, 2, \cdots, m \end{cases}$$

Let $\{\phi_i\}$ be the set of the standard bases of V_h. Let $\{\phi_i^k\}$ be the subset of $\{\phi_i\}$ related to Ω_k, $(k = 1, 2)$. Let $\{\phi_i^\Gamma\}$ be the subset of $\{\phi_i\}$ related to Γ. Suppose that the nodes are ordered in the way: first those on $\Omega_k, k = 1, 2$ and then $\{\xi_j\}_{j=1}^m$. Let

$$u_h = \sum_i \alpha_i^1\phi_i^1 + \sum_j \alpha_j^2\phi_j^2 + \sum_{k=1}^m \alpha_k^\Gamma\phi_k^\Gamma$$

be the solution of (3.3.1). Denote

$$U_1 = (\alpha_i^1),\ U_2 = (\alpha_j^2),\ U_3 = (\alpha_k^\Gamma).$$

(3.3.1) can then be rewritten as

$$KU = \begin{bmatrix} K_{11} & 0 & K_{13} \\ 0 & K_{22} & K_{23} \\ K_{13}^T & K_{23}^T & K_{33} \end{bmatrix} \begin{bmatrix} U_1 \\ U_2 \\ U_3 \end{bmatrix} = \begin{bmatrix} b_1 \\ b_2 \\ b_3 \end{bmatrix}, \tag{3.3.8}$$

where

$$K = \Big(A(\phi_i,\phi_j)\Big), \qquad K_{kk} = \Big(A(\phi_i^k,\phi_j^k)\Big), \ K_{k3} = \Big(A(\phi_i^k,\phi_j^\Gamma)\Big),$$
$$K_{33}^{(k)} = \Big(A_k(\phi_i^\Gamma,\phi_j^\Gamma)\Big), \ K_{33} = K_{33}^{(1)} + K_{33}^{(2)}, \quad k=1,2.$$

By block Gaussian elimination, (3.3.8) reduces to

$$S^h U_3 = \tilde{b}_3, \tag{3.3.9}$$

where, $\tilde{b}_3 = b_3 - K_{13}^T K_{11}^{-1} b_1 - K_{23}^T K_{22}^{-1} b_2,$

$$S^h = K_{33} - K_{13}^T K_{11}^{-1} K_{13} - K_{23}^T K_{22}^{-1} K_{23} = S_1^h + S_2^h, \tag{3.3.10}$$

$$S_k^h = K_{33}^{(k)} - K_{k3}^T K_{kk}^{-1} K_{k3}, \quad k=1,2. \tag{3.3.11}$$

Note that in the equality

$$\begin{bmatrix} K_{kk} & K_{k3} \\ K_{k3}^T & K_{33}^{(k)} \end{bmatrix} \left\{ \begin{bmatrix} K_{kk} & K_{k3} \\ 0 & I \end{bmatrix}^{-1} \begin{bmatrix} 0 \\ \lambda \end{bmatrix} \right\} = \begin{bmatrix} 0 \\ S_k^h \lambda \end{bmatrix}, \quad k=1,2, \tag{3.3.12}$$

the first matrix corresponds to $\dfrac{\partial}{\partial \mathbf{n}_k}$, while the part in the braces corresponds to the harmonic extension function $E_k^h \lambda$ for $\lambda \in \Re^m$ (cf.[83]). Therefore, S_k^h and S^h are, respectively, the discrete counterparts of S_k and S. S^h and S_k^h are called, correspondingly, the capacitance matrices[74] of Ω and Ω_k with regard to Γ. Sometimes, S^h is called the Schur complement. S^h and S_k^h are symmetric, positive definite matrices and spectrally equivalent, which is implied by

Theorem 3.3.3. *Let Ω_h be quasi-uniform. Then there exist two constants c and C, independent of h, such that*

$$c\lambda^T S_k^h \lambda \le \lambda^T S^h \lambda \le C \lambda^T S_k^h \lambda, \ \forall\, \lambda \in \Re^m, \ k=1,2. \tag{3.3.13}$$

Proof. It follows from the definitions of E_k^h and S_k^h that

$$\lambda^T S_k^h \mu = A_k(E_k^h \lambda, E_k^h \mu), \ k=1,2, \ \forall\, \lambda,\mu \in \Re^m \tag{3.3.14}$$

Therefore, by Theorem 3.3.1, we eventually get (3.3.13). □

For $k=1,2$, define $D_k^h : V_h \to \Re^m$ as

$$\forall\, v \in V_h, \ D_k^h v \in \Re^m : \ \Big(D_k^h v\Big)(j) = A_k(v, \phi_j^\Gamma) - (f, \phi_j^\Gamma)_k, \ j = 1,2,\cdots,m.$$

$\forall\, v \in V_h$, let $v^k \in V_h^k$, such that

$$v^k(x) = \begin{cases} v(x), & \forall \text{ node } x \in \overline{\Omega}_k \\ 0, & \forall \text{ node } x \in \Omega \backslash \overline{\Omega}_k \end{cases}$$

With the above notations, definitions and theorems, we can state and analyse all the algorithms in two–subdomain nonoverlap case, which are presented in the following sections.

3.4 Iterative substructuring algorithms

Algorithm 3.4.1.[65] (Marini–Quarteroni scheme)

Step 1. Let $g^0 \in \Re^m$ be any initial guess, and set n:=1.

Step 2. Solve the Dirichlet subproblem posed on Ω_1:

$$\begin{cases} u_1^n \in V_h^1 \\ A_1(u_1^n, v) = (f, v)_1, \ \forall\, v \in V_h^{1,0} \\ r_0^h u_1^n = g^{n-1} \end{cases}$$

Step 3. Solve the Dirichlet–Neumann subproblem set on Ω_2:

$$\begin{cases} u_2^n \in V_h^2 \\ A_2(u_2^n, v) = (f, v)_2, \ \forall\, v \in V_h^{2,0} \\ A_2(u_2^n, \rho_2\lambda) = (f, \rho_1\lambda)_1 + (f, \rho_2\lambda)_2 - A_1(u_1^n, \rho_1\lambda), \ \forall\, \lambda \in \Re^m \end{cases}$$

Step 4. Choose the relaxation factor $\theta_n \in (0, 1)$. Calculate

$$g^n = \theta_n r_0^h u_2^n + (1 - \theta_n) g^{n-1}.$$

Set n:=n+1, return to Step 2 until some reasonable stopping criterion is satisfied.

Theorem 3.4.1. *Let u_h be the solution of (3.3.1). u_1^n, u_2^n and g^n are generated by Algorithm 3.4.1. Denote $\varepsilon_k^n = u_k^n - u_h^k$ $(k = 1, 2)$, $\mu^n = g^n - r_0^h u_h = g^n - U_3$. Then*

1)

$$\frac{1}{\tau} A_1(\varepsilon_1^n, \varepsilon_1^n) \le A_2(\varepsilon_2^n, \varepsilon_2^n) \le \sigma A_1(\varepsilon_1^n, \varepsilon_1^n). \quad (3.4.1)$$

2) There exists a constant $\theta^* \in (0, 1]$, such that

$$A_1(\varepsilon_1^{n+1}, \varepsilon_1^{n+1}) \le \kappa(\theta_n) A_1(\varepsilon_1^n, \varepsilon_1^n), \quad (3.4.2)$$

where $\kappa(\theta_n) < 1, \ \forall\, \theta_n \in (0, \theta^*)$.

3) There exists the optimal relaxation factor θ^{opt}, such that

$$\kappa(\theta^{opt}) = \min_{\theta \in (0, \theta^*)} \kappa(\theta).$$

4) Algorithm 3.4.1 is equivalent to the following iterative method for the capacitance equation (3.3.9):

$$S_2^h(g^{n+1} - g^n) = \theta_{n+1}(\tilde{b}_3 - S^h g^n). \quad (3.4.3)$$

Proof. It is easy to see that $\varepsilon_k^{n+1} \in V_h^k$ satisfies

$$\begin{cases} A_1(\varepsilon_1^{n+1}, v) = 0, \ \forall\ v \in V_h^{1,0} \\ r_0^h \varepsilon_1^{n+1} = \mu^n \end{cases} \tag{3.4.4}$$

$$\begin{cases} A_2(\varepsilon_2^{n+1}, v) = 0, \ \forall\ v \in V_h^{2,0} \\ A_2(\varepsilon_2^{n+1}, \rho_2\lambda) = -A_1(\varepsilon_1^{n+1}, \rho_1\lambda), \ \forall\ \lambda \in \Re^m \end{cases} \tag{3.4.5}$$

$$\mu^{n+1} = \theta_{n+1} r_0^h \varepsilon_2^{n+1} + (1 - \theta_{n+1})\mu^n. \tag{3.4.6}$$

(3.4.4) and (3.4.5) yield

$$A_2(\varepsilon_2^{n+1}, v) = -A_1(\varepsilon_1^{n+1}, E_1^h r_0^h v), \ \forall\ v \in V_h^2 \tag{3.4.7}$$

which indicates that $\varepsilon_2^{n+1} = T_2^h r_0^h \varepsilon_1^{n+1}$. By (3.3.6), we get (3.4.1).

On the other hand, it follows from (3.4.4) and (3.4.6) that

$$\varepsilon_1^{n+1} = \theta_n E_1^h r_0^h \varepsilon_2^n + (1 - \theta_n)\varepsilon_1^n.$$

Hence

$$\begin{aligned} A_1(\varepsilon_1^{n+1}, \varepsilon_1^{n+1}) = \ & \theta_n^2 A_1(E_1^h r_0^h \varepsilon_2^n, E_1^h r_0^h \varepsilon_2^n) + (1 - \theta_n)^2 A_1(\varepsilon_1^n, \varepsilon_1^n) \\ & + 2\theta_n(1 - \theta_n) A_1(E_1^h r_0^h \varepsilon_2^n, \varepsilon_1^n). \end{aligned} \tag{3.4.8}$$

By (3.3.2) and (3.4.1), we obtain

$$A_1(E_1^h r_0^h \varepsilon_2^n, E_1^h r_0^h \varepsilon_2^n) \le \sigma A_2(E_2^h r_0^h \varepsilon_2^n, E_2^h r_0^h \varepsilon_2^n) \le \sigma^2 A_1(\varepsilon_1^n, \varepsilon_1^n). \tag{3.4.9}$$

The substitution of v in (3.4.7) with ε_2^{n+1} gives

$$A_1(E_1^h r_0^h \varepsilon_2^n, \varepsilon_1^n) = -A_2(\varepsilon_2^n, \varepsilon_2^n). \tag{3.4.10}$$

(3.4.1) implies that

$$A_2(\varepsilon_2^n, \varepsilon_2^n) \ge \frac{1}{\tau} A_1(\varepsilon_1^n, \varepsilon_1^n). \tag{3.4.11}$$

If $0 < \theta_n < 1$, then (3.4.2) results from the substitution of (3.4.9), (3.4.10) and (3.4.11) into (3.4.8). Furthermore, $\kappa(\theta_n)$ has the following expression

$$\kappa(\theta_n) = \frac{1}{\tau}\Big(\theta_n^2(\sigma^2\tau + \tau + 2) - 2\theta_n(\tau + 1) + \tau\Big). \tag{3.4.12}$$

An elementary calculation indicates that if and only if

$$0 < \theta_n < \theta^* = \min\Big(1, \frac{2(\tau + 1)}{\sigma^2\tau + \tau + 2}\Big),$$

then $0 \le \kappa(\theta_n) < 1$. Besides, the optimal relaxation factor

$$\theta^{opt} = \frac{\tau + 1}{\sigma^2\tau + \tau + 2}, \tag{3.4.13}$$

satisfies

$$\kappa(\theta^{opt}) = \frac{\sigma^2\tau^2 - 1}{\tau(\sigma^2\tau + \tau + 2)} = \min_{\theta\in(0,\theta^*)} \kappa(\theta). \tag{3.4.14}$$

By (3.4.4), (3.4.5) and (3.3.12), we have $r_0^h\varepsilon_2^{n+1} = -[S_2^h]^{-1}S_1^h\mu^n$, which combining (3.4.6) yields (3.4.3). □

Algorithm 3.4.2.[9] (Trace averaging method)

Step 1. Let $g^0 \in \Re^m$ be any initial guess. Set n:=1. Let $\rho \in (0,1)$ be the given relaxation factor.

Step 2. For $k = 1, 2$, solve in parallel $u_k^n \in V_h^k$, such that

$$\begin{cases} A_k(u_k^n, v) = (f, v)_k, \ \forall\ v \in V_h^{k,0} \\ r_0^h u_k^n = g^{n-1} \end{cases}$$

Step 3. For $k = 1, 2$, solve in parallel $w_k^n \in V_h^k$, such that

$$A_k(w_k^n, v) = \frac{1}{2}\sum_{i=1}^{2}\Big(A_i(u_i^n, \rho_i r_0^h v) - (f, \rho_i r_0^h v)_i\Big),\ \forall\ v \in V_h^k$$

Step 4. Calculate $g^n = g^{n-1} - \rho(r_0^h w_1^n + r_0^h w_2^n)$, set n:=n+1 and then go to Step 2.

Now, lets investigate Algorithm 3.4.2. Suppose that u_h is the solution of (3.3.1) and u_1^n, u_2^n and g^n are generated by Algorithm 3.4.2. Then

$$\varepsilon_k^n = u_k^n - u_h^k,\ k = 1, 2,\ \mu^n = g^n - r_0^h u_h = g^n - U_3$$

satisfy

$$\begin{cases} A_k(\varepsilon_k^n, v) = 0, \ \forall\ v \in V_h^{k,0} \\ r_0^h \varepsilon_k^n = \mu^{n-1} \end{cases} \tag{3.4.15}$$

$$A_k(w_k^n, v) = \frac{1}{2}\sum_{i=1}^{2} A_i(\varepsilon_i^n, \rho_i r_0^h v),\ \forall\ v \in V_h^k \tag{3.4.16}$$

$$\mu^{n+1} = \mu^n - \rho(r_0^h w_1^n + r_0^h w_2^n). \tag{3.4.17}$$

(3.4.15) and (3.4.16) give

$$A_k(w_k^n, v) = \frac{1}{2}\sum_{i=1}^{2} A_i(\varepsilon_i^n, E_i^h r_0^h v),\ \forall\ v \in V_h^k \tag{3.4.18}$$

It follows from (3.4.15), (3.4.17) and (3.4.18) that

$$\varepsilon_1^{n+1} = \varepsilon_1^n - \rho(w_1^n + E_1^h r_0^h w_2^n), \quad \varepsilon_2^{n+1} = \varepsilon_2^n - \rho(w_2^n + E_2^h r_0^h w_1^n). \tag{3.4.19}$$

By (3.4.15) and the substitution of v in (3.4.18) with ε_k^n, we get

$$\sum_{i=1}^{2} A_i(\varepsilon_i^n, \varepsilon_i^n) = \sum_{i=1}^{2} A_i(w_i^n, \varepsilon_i^n) \le \Big(\sum_{i=1}^{2} A_i(\varepsilon_i^n, \varepsilon_i^n)\Big)^{\frac{1}{2}} \Big(\sum_{i=1}^{2} A_i(w_i^n, w_i^n)\Big)^{\frac{1}{2}}.$$

Hence

$$\sum_{i=1}^{2} A_i(\varepsilon_i^n, \varepsilon_i^n) \le \sum_{i=1}^{2} A_i(w_i^n, w_i^n). \tag{3.4.20}$$

Taking $v = w_k^n$ in (3.4.18) yields

$$\sum_{i=1}^{2} A_i(w_i^n, w_i^n) = \frac{1}{2} A_1(\varepsilon_1^n, w_1^n + E_1^h r_0^h w_2^n) + \frac{1}{2} A_2(\varepsilon_2^n, w_2^n + E_2^h r_0^h w_1^n). \tag{3.4.21}$$

The substitution of v in (3.4.18) with $E_1^h r_0^h w_2^n$ for $k=1$ and $E_2^h r_0^h w_1^n$ for $k=2$ gives

$$A_1(w_1^n, E_1^h r_0^h w_2^n) = \frac{1}{2}\Big(A_1(\varepsilon_1^n, E_1^h r_0^h w_2^n) + A_2(\varepsilon_2^n, w_2^n)\Big),$$

$$A_2(w_2^n, E_2^h r_0^h w_1^n) = \frac{1}{2}\Big(A_2(\varepsilon_2^n, E_2^h r_0^h w_1^n) + A_1(\varepsilon_1^n, w_1^n)\Big).$$

Summing up the above two equalities and applying (3.4.21), we obtain

$$\sum_{i=1}^{2} A_i(w_i^n, w_i^n) = A_1(w_1^n, E_1^h r_0^h w_2^n) + A_2(w_2^n, E_2^h r_0^h w_1^n). \tag{3.4.22}$$

It follows from (3.4.19), (3.4.21), (3.4.22) and Theorem 3.3.1 that

$$\begin{aligned}\sum_{i=1}^{2} A_i(\varepsilon_i^{n+1}, \varepsilon_i^{n+1}) &= \sum_{i=1}^{2} A_i(\varepsilon_i^n, \varepsilon_i^n) - 4\rho \sum_{i=1}^{2} A_i(w_i^n, w_i^n) + \rho^2\Big\{3\sum_{i=1}^{2} A_i(w_i^n, w_i^n) \\ &\quad + A_1(E_1^h r_0^h w_2^n, E_1^h r_0^h w_2^n) + A_2(E_2^h r_0^h w_1^n, E_2^h r_0^h w_1^n)\Big\} \\ &\le \sum_{i=1}^{2} A_i(\varepsilon_i^n, \varepsilon_i^n) - \Big(4\rho - (3+\beta)\rho^2\Big) \sum_{i=1}^{2} A_i(w_i^n, w_i^n),\end{aligned}$$

where $\beta = \max(\sigma, \tau)$. Furthermore, by (3.4.20), we see that if $4\rho - (3+\beta)\rho^2 \ge 0$, then

$$\sum_{i=1}^{2} A_i(\varepsilon_i^{n+1}, \varepsilon_i^{n+1}) \le \eta(\rho) \sum_{i=1}^{2} A_i(\varepsilon_i^n, \varepsilon_i^n),$$

where $\eta(\rho) = 1 - \Big(4 - (3+\beta)\rho\Big)\rho$. Obviously, we have

$$0 \le \eta(\rho) < 1, \quad \forall\, \rho \in (0, \rho^*) = (0, \frac{4}{3+\beta})$$

$$\eta(\rho^{opt}) = \min_{\rho\in(0,\rho^*)} \eta(\rho) = 1 - \frac{4}{3+\beta}, \quad \text{where } \rho^{opt} = \frac{2}{3+\beta}. \tag{3.4.23}$$

By now, we get the following theorem.

Theorem 3.4.2. 1) *Let u_h be the solution of (3.3.1). u_1^n and u_2^n are generated by Algorithm 3.4.2. Let ε^n be the iterative error, s.t. $\varepsilon^n = u_i^n - u_h^i$ on Ω_i. Then, there exists a constant $\rho^* \in (0,1)$, such that*

$$A(\varepsilon^{n+1}, \varepsilon^{n+1}) \le \eta(\rho) A(\varepsilon^n, \varepsilon^n), \tag{3.4.24}$$

and $\eta(\rho) < 1$, $\forall\ \rho \in (0, \rho^*)$. Besides, the optimal relaxation factor ρ^{opt} exists.

2) Algorithm 4.2 is equivalent to the following iterative method for (3.3.9)

$$g^{n+1} = g^n + \frac{1}{2}\rho\Big([S_1^h]^{-1} + [S_2^h]^{-1}\Big)(\tilde{b}_3 - S^h g^n). \tag{3.4.25}$$

Algorithm 3.4.3.[89] (Parallel iterative method)

Step 1. Let $g^n \in \Re^m$ be any initial guess. Set n:=0. Choose the relaxation factors $\theta_1, \theta_2 \in (0,1)$.

Step 2. For $k = 1, 2$, solve in parallel the Dirichlet subproblems on Ω_k

$$\begin{cases} u_k^{2n} \in V_h^k \\ A_k(u_k^{2n}, v) = (f, v)_k, \ \forall\ v \in V_h^{k,0} \\ r_0^h u_k^{2n} = g^n \end{cases}$$

Step 3. Compute

$$d^n = \theta_1 D_1^h u_1^{2n} - (1-\theta_1) D_2^h u_2^{2n}.$$

Step 4. For $k = 1, 2$, solve simultaneously the mixed subproblems on Ω_k

$$\begin{cases} u_k^{2n+1} \in V_h^k \\ A_k(u_k^{2n+1}, v) = (f, v)_k, \ \forall\ v \in V_h^{k,0} \\ A_k(u_k^{2n+1}, \phi_i^\Gamma) = (f, \phi_i^\Gamma)_k + (-1)^{k+1}\Big(d^n\Big)(i), \ i = 1, 2, \cdots, m \end{cases}$$

Step 5. Calculate

$$g^{n+1} = \theta_2 r_0^h u_1^{2n+1} + (1-\theta_2) r_0^h u_2^{n+1},$$

set n:=n+1, and go back to Step 2 for the next iteration.

Theorem 3.4.3. *1) Let ε^n be the error of Algorithm 3.4.3, where $\varepsilon^n = u_i^n - u_h^i$ on Ω_i. Then, if*

$$\theta_1 \in (\theta_1^*, 1) = (\frac{\tau^2\sigma - \sigma}{\tau^2\sigma + \sigma + 2}, 1), \quad \theta_2 \in (\theta_2^*, 1) = (\frac{\sigma^2\tau - \tau}{\sigma^2\tau + \tau + 2}, 1),$$

then

$$\kappa_1(\theta_1) = \frac{1}{\sigma}\Big\{(1-\theta_1)^2(\sigma\tau^2+\sigma+2) - 2(1-\theta_1)(\sigma+1)+\sigma\Big\} \in [0,1),$$

$$\kappa_2(\theta_2) = \frac{1}{\tau}\Big\{(1-\theta_2)^2(\tau\sigma^2+\tau+2) - 2(1-\theta_2)(\tau+1)+\tau\Big\} \in [0,1),$$

satisfy

$$A(\varepsilon^{n+2},\varepsilon^{n+2}) \le \kappa_1(\theta_1)\kappa_2(\theta_2)A(\varepsilon^n,\varepsilon^n).$$

2)

$$\theta_1^{opt} = \frac{\tau^2\sigma+1}{\tau^2\sigma+\sigma+2}, \quad \theta_2^{opt} = \frac{\sigma^2\tau+1}{\sigma^2\tau+\tau+2}, \tag{3.4.26}$$

satisfy

$$\begin{cases} \kappa_1(\theta_1^{opt}) = \dfrac{\sigma^2\tau^2-1}{\sigma(\tau^2\sigma+\sigma+2)} = \min\limits_{\theta\in(\theta_1^*,1)} \kappa_1(\theta) \\ \kappa_2(\theta_2^{opt}) = \dfrac{\sigma^2\tau^2-1}{\tau(\sigma^2\tau+\tau+2)} = \min\limits_{\theta\in(\theta_2^*,1)} \kappa_2(\theta) \end{cases} \tag{3.4.27}$$

3) Algorithm 3.4.3 is equivalent to the following iterative method for (3.3.9)

$$g^{n+1} = g^n + \Big\{\theta_2(1-\theta_1)[S_1^h]^{-1} + \theta_1(1-\theta_2)[S_2^h]^{-1}\Big\}(\tilde{b}_3 - S^h g^n). \tag{3.4.28}$$

As before, Theorem 3.4.3 can be proved by Theorem 3.3.1 and Corrollary 3.3.2. For details, refer to [41].

On the whole, the above iterative algorithms converge geometrically, independently of h. Besides, the optimal relaxation factors exist. In the special case that the equation (3.2.1) and the domain Ω are symmetric in respect of Γ, $\sigma = \tau = 1$, then $\kappa(\theta^{opt}) = 0$ in (3.4.14), $\eta(\rho^{opt}) = 0$ in (3.4.23), $\kappa_1(\theta_1^{opt}) = \kappa_2(\theta_2^{opt}) = 0$ in (3.4.27), which imply that only one iteration is needed to obtained the solution of (3.3.1). And, they are equivalent to the simple iterative method for the capacitance equation (3.3.9) which have been preconditioned, respectively, by

$$\frac{1}{\theta}S_2^h, \ \frac{2}{\rho}\Big([S_1^h]^{-1}+[S_2^h]^{-1}\Big)^{-1}, \ \Big\{\theta_2(1-\theta_1)[S_1^h]^{-1} + \theta_1(1-\theta_2)[S_2^h]^{-1}\Big\}^{-1}. \tag{3.4.29}$$

Theorem 3.3.3 shows that all the preconditioners in (3.4.29) are equivalent to S^h and the relaxation factors can be selected so that the spectral radii of the iterative matrices of (3.4.3), (3.4.25) and (3.4.28) are less than one.

3.5 Substructuring preconditioners

Generally, the capacitance matrix S^h is dense and its condition number is $O(h^{-1})$ (cf.[25],(3.3.3),(3.3.14)). Hence, to solve (3.3.9), the iterative methods are peferable to the direct methods. It is well–known that the simple iterative method converges much slowly than the conjugate gradient method (CG). Consequently, we may use one of the preconditioners in (3.4.29) or any others implied in the iterative methods just like Algorithm 3.4.1—3.4.3 to accelerate the CG method for solving (3.3.9), i.e., to apply the preconditioned conjugate gradient method (PCG) to solving (3.3.9).

Let $S_0^h(\Gamma) \subset H_0^1(\Gamma)$ be the piecewise linear continuous finite element space with $\{\xi_j\}_{j=1}^m$ as the interpolation points. We note that the discrete operator l_0 defined on $S_0^h(\Gamma)$ by

$$\langle l_0 v, w\rangle = \int_\Gamma \frac{dv}{ds}\frac{dw}{ds} ds, \quad \forall\, w \in S_0^h(\Gamma)$$

is a finite–dimensional approximation to the Laplace operator $-\dfrac{d^2}{ds^2}$. Here, s is the arc length along Γ. Let $l_0^{\frac{1}{2}}$ be the positive square root of the symmetric, positive definite matrix l_0. It follows from [12] and Theorem 3.3.1 that there exists two constants c and C , independent of h, such that

$$c\lambda^T S^h \lambda \le \langle l_0^{\frac{1}{2}} \lambda_h, \lambda_h\rangle \le C\lambda^T S^h\lambda, \quad \forall\, \lambda \in \Re^m \tag{3.5.1}$$

where λ_h is determined uniquely by λ as that in Theorem 3.3.1.

By now, there has been some investigation on the $m \times m$ matrix $l_0^{\frac{1}{2}}$. For $j = 1, 2, \cdots, m$, denote

$$E_j = \sqrt{\frac{2}{m+1}}\Big(\sin\frac{j\pi}{m+1}, \sin\frac{2j\pi}{m+1}, \cdots, \sin\frac{mj\pi}{m+1}\Big)^T, \; E = (E_1, E_2, \cdots, E_m),$$

$$\xi_j = 4\sin^2\frac{j\pi}{2(m+1)}, \; \alpha_j^D = 2\sqrt{\xi_j}, \; \alpha_j^G = 2\sqrt{\xi_j + \xi_j^2/4}, \; \alpha_j^J = \sqrt{\xi_j(6-\xi_j)/6},$$

$$D = E\ diag(\alpha_1^D, \alpha_2^D, \cdots, \alpha_m^D)\ E^T, \tag{3.5.2}$$

$$G = E\ diag(\alpha_1^G, \alpha_2^G, \cdots, \alpha_m^G)\ E^T, \tag{3.5.3}$$

$$J = E\ diag(\alpha_1^J, \alpha_2^J, \cdots, \alpha_m^J)\ E^T. \tag{3.5.4}$$

We point out that $l_0^{\frac{1}{2}}$ becomes J when $\{\xi_j\}_{j=1}^m$ are equally spaced on Γ (cf.[12]).

Theorem 3.5.1.[12,30,39] *Let S^h, D, G and J be defined by (3.3.10), (3.5.2), (3.5.3) and (3.5.4), respectively. If Ω_h is quasi–uniform, then D, G and J are spectrally equivalent to S^h.*

The above theorem indicates that even if S^h is the capacitance matrix in the nonconforming cases, D, G and J are still efficient preconditioners to apply PCG method to solving (3.3.9). In each iteration, $D^{-1}g, G^{-1}g$ and $J^{-1}g$ can be obtained by FFT, while the number of iterations is independent of the size of the problems.

By Theorem 3.3.3, it is easy to see that

$$Q = \begin{bmatrix} K_{11} & 0 & K_{13} \\ 0 & K_{22} & K_{23} \\ K_{13}^T & K_{23}^T & K_{33}^{(1)} + K_{23}^T K_{22}^{-1} K_{23} \end{bmatrix} \tag{3.5.5}$$

is spectrally equivalent to the stiff matrix K in (3.3.8), thus Q can be used as a preconditioner to apply PCG method to solving (3.3.8). In each iteration, $Q^{-1}g$ can be calculated through first solving simultaneously the Dirichlet subproblems on Ω_1 and Ω_2 and then solving a Dirichlet–Neumann subproblem on Ω_1. Consequently, it is equivalent to the so–called substructure partitioned method[11]. In fact, Q is the matrix representation of the preconditioner developed in [11].

By now, we have surveyed all existing algorithms in two–subdomain nonoverlap cases.

3.6 Numerical experiments

Example 3.6.1. We use the Crouzeix–Raviart elements to discretize

$$\begin{cases} \Delta u = f, & \text{in } \Omega = (0,2)\times(0,1)\cup(0,1)\times[1,2) \\ u = 0, & \text{on } \partial\Omega \end{cases} \tag{3.6.1}$$

Ω is divided into $\Omega_1 = (0,1)\times(1,2)$ and $\Omega_2 = (0,2)\times(0,1)$. See Fig.3.6.1. We triangulate Ω to get the fine mesh Ω_h so that each element $e \in \Omega_h$ is an isosceles right triangle with h as its diameter. In Ω_1 and Ω_2, there are, respectively, 16, 32 elements, 28, 54 nodes for $h = 1/2$, 64,128 elements, 104, 204 nodes for $h = 1/4$, 256, 512 elements, 400, 792 nodes for $h = 1/8$. In the following tables, n is the number of iterations, and $n = 0$ represents the initial situation. ε^n is the error after n iterations.

$$\|\varepsilon^n\|_A = \sqrt{A(\varepsilon^n,\varepsilon^n)}, \qquad \rho_n = \sqrt[n]{\|\varepsilon^n\|_A/\|\varepsilon^0\|_A},$$
$$\|\varepsilon^n\|_\infty = \|\varepsilon^n\|_{L^\infty(\Omega)}, \qquad \delta_n = \sqrt[n]{\|\varepsilon^n\|_\infty/\|\varepsilon^0\|_\infty}.$$

We apply Algorithm 3.4.1 to solving (3.6.1), where a procedure is built up to generate a sequence of nonconforming discrete harmonic functions on Ω_1 and Ω_2 with the same nodal values on Γ. This allows us to compute, at each iteration, two constants σ_n, τ_n suggested by (3.3.2), which combining (3.4.13) leads to the sequence of approximate values θ_n of the optimal relaxation factor θ^{opt}. We point out that the evaluation of θ_n doesn't require the solution of any additional problem in our algorithm (for details, cf.[65]). The main experimental results are listed in Table 3.6.1 and Table 3.6.2, which coincide with our theoretical analysis.

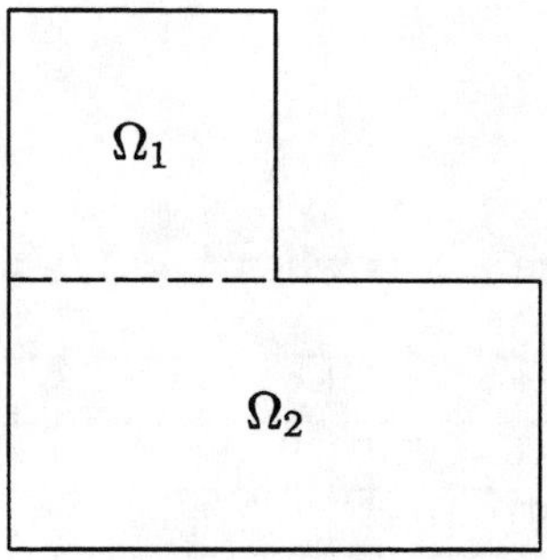

Fig.3.6.1

Example 3.6.2. We use the Carey membrane elements to discretize

$$\begin{cases} -\triangle u + u = f, & \text{in } \Omega \\ u = 0, & \text{on } \partial\Omega \end{cases} \tag{3.6.2}$$

Here, Ω and its subdivision Ω_1 and Ω_2, and the fine mesh Ω_h are the same as those in Example 3.6.1. In Ω_1, Ω_2, there are, respectively, 29, 55 nodes for $h = 1/2$, 105, 205 nodes for $h = 1/4$, 401, 793 nodes for $h = 1/8$. As before, Algorithm 3.4.1 is applied to solving (3.6.2) adaptively, since the intermediate values of the iterations can be used to calculate the approximate value of the optimal relaxation factor θ^{opt}. The numerical results are listed in Table 3.6.3 and Table 3.6.4, which show an impressive reduction of the initial error after very few iterations.

Table 3.6.1: ρ_n, δ_n vs. h for Example 3.6.1

	n	2	4	6	8	10
ρ_n	$h = 1/2$	0.0179	0.0187	0.0188	0.0189	0.0190
	$h = 1/4$	0.0020	0.0119	0.0196	0.0243	0.0275
	$h = 1/8$	0.0045	0.0205	0.0309	0.0370	0.0409
δ_n	$h = 1/2$	0.1285	0.1345	0.1359	0.1365	0.1369
	$h = 1/4$	0.0414	0.0985	0.1308	0.1481	0.1587
	$h = 1/8$	0.0611	0.1285	0.1633	0.1815	0.1926

Table 3.6.2: The errors when $h = 1/8$ for Example 3.6.1

n	2	4	6	8	10
$\|\varepsilon^n\|_A$	$0.335 \cdot 10^1$	$0.646 \cdot 10^{-2}$	$0.212 \cdot 10^{-4}$	$0.708 \cdot 10^{-7}$	$0.242 \cdot 10^{-9}$
$\|\varepsilon^n\|_\infty$	$0.112 \cdot 10^1$	$0.391 \cdot 10^{-1}$	$0.214 \cdot 10^{-2}$	$0.119 \cdot 10^{-3}$	$0.669 \cdot 10^{-5}$

Table 3.6.3: ρ_n, δ_n vs. h for Example 3.6.2

n		2	4	5	6	7
ρ_n	$h = 1/2$	0.0002	0.0021	0.0041	0.0019	0.0053
	$h = 1/4$	0.0001	0.0005	0.0007	0.0015	0.0036
	$h = 1/8$	0.0002	0.0016	0.0021	0.0028	0.0054
δ_n	$h = 1/2$	0.0035	0.0277	0.0324	0.0428	0.0441
	$h = 1/4$	0.0087	0.0209	0.0255	0.0395	0.0604
	$h = 1/8$	0.0091	0.0364	0.0442	0.0559	0.0763

Table 3.6.4: The errors when $h = 1/8$ for Example 3.6.2

n	2	4	5	6	7
$\|\varepsilon^n\|_A$	$0.760 \cdot 10^{-1}$	$0.164 \cdot 10^{-5}$	$0.780 \cdot 10^{-8}$	$0.687 \cdot 10^{-10}$	$0.897 \cdot 10^{-11}$
$\|\varepsilon^n\|_\infty$	$0.105 \cdot 10^0$	$0.555 \cdot 10^{-3}$	$0.441 \cdot 10^{-4}$	$0.628 \cdot 10^{-5}$	$0.226 \cdot 10^{-5}$

Chapter 4

An Iterative Algorithm with Multi–subdomains for the First Kind Nonconforming Elements

4.1 Introduction

From this chapter to Chapter 8, we consider the substructuring algorithms with multi–subdomains. It is rather complicated in the case of multi-subdomains with the internal cross points. A cross point is defined to be the common boundary point of more than two subdomains. With the techniques of the separation of the internal cross points from other mesh nodes, Bramble[12–15] et al., constructed different preconditioners for the algebraic system of equations which arise from the conforming finite element discretizations of the following selfadjoint elliptic problems:

$$u \in H_0^1(\Omega): \; a(u,v) = (f,v), \quad \forall\, v \in H_0^1(\Omega), \tag{4.1.1}$$

where

$$a(u,v) = \int_\Omega \left[\sum_{i,j=1}^{2} a_{ij}(x) \frac{\partial u}{\partial x_i} \frac{\partial v}{\partial x_j} + a_0(x) uv \right], \quad (f,v) = \int_\Omega fv, \tag{4.1.2}$$

$\Omega \subseteq \Re^2$ is a bounded polygonal open domain, $f \in H^{-1}(\Omega)$, $a_0(x) \geq 0$, $a_{ij}(x)$ $(i,j = 1,2)$, $a_0(x)$ are piecewise smooth and bounded functions in Ω, (a_{ij}) is a symmetric, uniformly positive definite matrix in Ω. All these preconditioners can be inversed easily in parallel and precondition the stiff matrix with the condition number no more than $O\big((1 + \ln \frac{H}{h})^2\big)$, where H, h are the subdomain diameter and the fine mesh parameter, respectively. Bourgat[9] introduced an iterative substructuring method with the trace averaging operator to deal with the internal cross points, and illustrated its efficiency in the conforming discrete case with plenty of numerical experiments. Later, Chu[24] gave the theoretical proof of its convergence.

The present chapter is concerned with the construction of efficient iterative schemes for solving (4.1.1) discretized by the nonconforming finite elements of the first kind. Let $\Omega_h = \{e\}$ be a quasi-uniform mesh of Ω, where h is the mesh parameter and e, a triangle or a quadrilateral, represents typical element in Ω_h. From Chapter 1, we know that the first kind nonconforming element space V_h can be written as

$$V_h = \{v: \; v|_e = \theta|_e + w|_e, \; \theta|_e \text{ is linear (bilinear)}\,; w(x) = 0, \; \forall \text{ node } x \in \overline{\Omega},$$

$$w|_e \text{ is a finite order polynomial, } \forall\, e \in \Omega_h\}$$

Here, a node $x \in \overline{\Omega}$ is defined to be the vertex of some $e \in \Omega_h$. Denote

$$V_h^0 = \{v \in V_h :\ v(x) = 0,\ \forall \text{ node } x \in \partial\Omega\},$$

$$A(u,v) = \sum_{e\subset\Omega} \int_e \left[\sum_{i,j=1}^{2} a_{ij} \frac{\partial u}{\partial x_i}\frac{\partial v}{\partial x_j} + a_0 uv \right].$$

Then, the nonconforming finite element discrete problem for (4.1.1) is

$$u_h \in V_h^0 :\ A(u_h, v_h) = (f, v_h),\ \forall\, v_h \in V_h^0 \tag{4.1.3}$$

The goal is to extend the trace averaging domain decomposition algorithm[9,29,37] to the solving of (4.1.3). Let $\{\Omega_i\}_{i=1}^N$ be a nonoverlap subdomain division of Ω. In each iteration of our algorithm, the Dirichlet subproblems set on subdomains Ω_i are solved simultaneously, which is then followed by solving simultaneously the Neumann subproblems posed on Ω_i, where the discrete trace averaging operators r_0^i play a key role in the interchange of information between subdomains. r_0^i is a matrix (cf. Sect.4.2), which is only related to the nodes on $\partial\Omega_i\backslash\partial\Omega$ and to the numbers of subdomains by which a node is shared. Therefore, the existence of the internal cross points doesn't affect the efficiency of our algorithm. It is proved to be geometrically convergent and is essentially equivalent to the simple iterative method applied to the preconditioned capacitance equation. Furthermore, the preconditioner implied in the iteration can be inversed easily in parallel and the condition number of the preconditioned capacitance matrix is bounded by $O\Big((1+\ln\frac{H}{h})(1+H^{-2}+\ln\frac{H}{h})\Big)$.

The remainder is organized as follows. The domain decomposition algorithm for (4.1.3) is described in Sect.4.2. Its convergence analysis is established in Sect.4.3, while its matrix analysis is presented in Sect.4.4.

4.2 Domain decomposition algorithm

In what follows, we assume that there exists a positive constant α s.t. for $a_0(x)$ of (4.1.2)

$$a_0(x) \geq \alpha,\quad \forall\, x \in \Omega$$

From [21,55,86], we know that there exists a unique solution u_h of (4.1.3) which converges to the solution u of (4.1.1).

Let $\{\Omega_i\}_{i=1}^N$ be a subdivision of Ω, which satisfies:

A$_1$. Ω_i is an open triangle or an open quadrilateral, $\bigcup_{i=1}^N \overline{\Omega}_i = \overline{\Omega}$;

A$_2$. $\{\Omega_i\}_{i=1}^N \triangleq \Omega_H$, which is supposed to be a quasi-uniform mesh of Ω with H as its mesh parameter. Generally, $H \gg h$, therefore, Ω_H, Ω_h are called the coarse mesh and the fine mesh, respectively;

A$_3$. Ω_H is compatible with the fine mesh Ω_h, i.e., $e \cap \Omega_i$ is either empty or e for all $e \in \Omega_h$ and $\Omega_i \in \Omega_H$.

Denote $\Gamma = \bigcup_{i=1}^{N} \partial\Omega_i \backslash \partial\Omega$. Let $\{\nu_j\}$ be the set of vertices of $\{\Omega_i\}_{i=1}^N$, i.e., the set of coarse mesh nodes. The open edge in Γ with endpoints ν_i and ν_j is denoted Γ_{ij}. Let $\{\chi_k\}_{k=1}^m$ be the set of the nodes on Γ (ordered in some way). Denote

$$I_i = \{k : \ \chi_k \in \partial\Omega_i \backslash \partial\Omega, \ k = 1, 2, \cdots, m\}.$$

Let m_i be the number of elements of I_i. For $i = 1, 2, \cdots, N$, the matrix $r_0^i \in \Re^{m\times m}$ satisfies

$$(r_0^i)_{lj} = \begin{cases} \frac{1}{k}, & l = j \in I_i, \ \chi_j \text{ is a common boundary point of } k \text{ subdomains} \\ 0, & l \notin I_i \ \text{ or } j \notin I_i \end{cases}$$

where $(r_0^i)_{lj}$ represents the (l,j)th element of r_0^i, $(l, j = 1, 2, \cdots, m)$. Obviously, we have

$$\sum_{i=1}^{N} r_0^i = I. \tag{4.2.1}$$

r_0^i is called *the discrete trace averaging operator*. The construction of r_0^i is an originality of the present approach. We define *the discrete trace operator* $r_0^h : V_h^0 \longrightarrow \Re^m$ as follows:

$$\forall\, v \in V_h^0, \quad r_0^h v \in \Re^m, \quad (r_0^h v)(j) = v(\chi_j), \quad j = 1, 2, \cdots, m.$$

And, we introduce the following notations:

$$V_h^i = \{v \in V_h^0 : \ v(x) = 0, \ \forall \text{ interpolation point } x \in \Omega\backslash\overline{\Omega}_i\},$$

$$V_h^{i,0} = \{v \in V_h^0 : \ v(x) = 0, \ \forall \text{ interpolation point } x \in \Omega\backslash\Omega_i\},$$

$$A_i(u, v) = \sum_{e\subset\Omega_i} \int_e \left[\sum_{k,j=1}^{2} a_{kj} \frac{\partial u}{\partial x_k} \frac{\partial v}{\partial x_j} + a_0 uv \right],$$

$$(f, v)_i = \int_{\Omega_i} fv, \quad i = 1, 2, \cdots, N.$$

Here, an interpolation point x is related to the definition of V_h.

Now we are in a position to describe the trace averaging domain decomposition method for the solving of the nonconforming discrete problem (4.1.3).

Algorithm 4.2.1.

Step 1. Choose arbitrary $\lambda^0 = (\lambda_1^0, \lambda_2^0, \cdots, \lambda_m^0)^T \in \Re^m$. Set n:=0.

Step 2. For $i = 1, 2, \cdots, N$, solve in parallel

$$\begin{cases} u_i^n \in V_h^i \\ A_i(u_i^n, \theta) = (f, \theta)_i, & \forall\, \theta \in V_h^{i,0} \\ u_i^n(\chi_j) = \lambda_j^n, & \forall \text{ node } \chi_j \in \partial\Omega_i \backslash \partial\Omega \\ u_i^n(x) = 0, & \forall \text{ node } x \in \partial\Omega_i \cap \partial\Omega \end{cases}$$

Step 3. For $i = 1, 2, \cdots, N$, solve in parallel

$$\begin{cases} \psi_i^n \in V_h^i \\ A_i(\psi_i^n, \theta) = \sum_{j=1}^{N} \left[A_j(u_j^n, Tr_j^{-1}(r_0^i r_0^h \theta)) - (f, Tr_j^{-1}(r_0^i r_0^h \theta))_j \right], \quad \forall\, \theta \in V_h^i \end{cases}$$

Step 4. Choose relaxation factor $0 < \rho < 1$ and compute

$$\lambda^{n+1} = \lambda^n - \rho \sum_{j=1}^{N} r_0^j r_0^h \psi_j^n.$$

Set $n := n+1$. Then, go back to Step 2 until some reasonable stopping criterion is satisfied.

Remark 4.2.1. In Algorithm 4.2.1, for $\lambda = (\lambda_1, \lambda_2, \cdots, \lambda_m)^T$, $Tr_j^{-1}(\lambda)$ means any element of the set

$$T_j(\lambda) = \{ w \in V_h^j : \ w(\chi_k) = \lambda_k, \ \forall \text{ node } \chi_k \in \partial\Omega_j \backslash \partial\Omega \}.$$

Its arbitrariness is implied by Step 2 and the fact that

$$w_1 - w_2 \in V_h^{j,0}, \ \forall\, w_1, w_2 \in T_j(\lambda)$$

4.3 Convergence analysis

For $i = 1, 2, \cdots, N$, denote: $|v|_{1,\Omega_i,h} = (\sum_{e \subset \Omega_i} |v|_{H^1(e)}^2)^{\frac{1}{2}}$. Define *the conforming interplation operator* $I_h : V_h \longrightarrow C(\overline{\Omega})$ as follows:

$$\forall\, v \in V_h, \ \ (I_h v)(x) = v(x), \ \ \forall \text{ node } x \in \overline{\Omega}$$

$$I_h v \text{ is piecewise linear (bilinear) on } \Omega_h.$$

It is the same as that in Sect.2.3 for the Carey elements, but different from that in Sect.2.3 for the Wilson elements and Wilson–like elements. I_h is introduced here just for theoretical analysis. It has the following property:

Lemma 4.3.1. *If Ω_h is quasi-uniform, then for $i = 1, 2, \cdots, N$*

$$\|v - I_h v\|_{L^\infty(\Omega_i)} \le c|v|_{1,\Omega_i,h}, \ \ \forall\, v \in V_h^i, \tag{4.3.1}$$

$$\|v - I_h v\|_{L^2(\Omega_i)} \le c\|v\|_{L^2(\Omega_i)}, \ \ \forall\, v \in V_h^i, \tag{4.3.2}$$

$$|v - I_h v|_{1,\Omega_i,h} \le c|v|_{1,\Omega_i,h}, \ \ \forall\, v \in V_h^i. \tag{4.3.3}$$

Proof. It follows from the interpolation theorem and an "inverse property" (see, e.g., [26]) implied by the quasi-uniformness of the mesh Ω_h that

$$\begin{aligned} \|v - I_h v\|_{L^\infty(\Omega_i)} &\le \max_{e \subset \Omega_i} \|v - I_h v\|_{L^\infty(e)} = \|v - I_h v\|_{L^\infty(e_0)} \\ &\le ch|v|_{H^2(e_0)} \le c|v|_{H^1(e_0)} \le c|v|_{1,\Omega_i,h}, \end{aligned}$$

where e_0 is some element in Ω_i. So (4.3.1) holds.

(4.3.2) and (4.3.3) can be established in the same manner. □

Lemma 4.3.2. *Let ν be a vertex of the subdomain Ω_i. If $w \in V_h^i$ satisfies*

$$\begin{cases} A_i(w,\theta) = 0, & \forall\, \theta \in V_h^{i,0} \\ w(x) = 0, & \forall \text{ node } x \in \partial\Omega_i,\ x \neq \nu \end{cases} \tag{4.3.4}$$

then

$$A_i(w,w) \leq c|w(\nu)|^2. \tag{4.3.5}$$

Proof. Let v be a piecewise linear continuous function defined on $\partial\Omega_i$ s.t.

$$v(x) = w(x), \quad \forall \text{ node } x \in \partial\Omega_i.$$

It follows from the proof of Theorem 2.4.3 that

$$A_i(w,w) \leq c||v||^2_{\frac{1}{2},\partial\Omega_i}. \tag{4.3.6}$$

Now, let's compute $||v||^2_{\frac{1}{2},\partial\Omega_i}$. Without loss of generality, we assume that Ω_i is a triangle with Γ_i $(i = 1,2,3)$ as its three edges, and ν is the common endpoint of Γ_1, Γ_3. It follows from the definition of the norm of the Sobolev space $H^{\frac{1}{2}}(\partial\Omega_i)$ (cf.[1,60]) that

$$||v||^2_{\frac{1}{2},\partial\Omega_i} = \int_{\partial\Omega_i}\int_{\partial\Omega_i} \frac{|v(x)-v(y)|^2}{|x-y|^2} ds(x)ds(y) + \frac{1}{|\partial\Omega_i|}\int_{\partial\Omega_i} v^2 ds = S_1 + S_2, \tag{4.3.7}$$

$$S_2 = \frac{1}{|\partial\Omega_i|}\int_{\partial\Omega_i} v^2 ds \leq \frac{1}{CH}\int_0^h ((1-\frac{1}{h}\xi)v(\nu))^2 d\xi \leq c|v(\nu)|^2, \tag{4.3.8}$$

$$S_1 = \sum_{k,j=1}^{3} \int_{\Gamma_k}\int_{\Gamma_j} \frac{|v(x)-v(y)|^2}{|x-y|^2} ds(x)ds(y). \tag{4.3.9}$$

Here, $|\partial\Omega_i|$ is the perimeter of Ω_i. By quasi-uniformness of Ω_H and the fact that $h \ll H$, simple calculation gives

$$\begin{aligned} &\int_{\Gamma_1}\int_{\Gamma_2} = \int_{\Gamma_2}\int_{\Gamma_1} \leq c|v(\nu)|^2, \quad \int_{\Gamma_1}\int_{\Gamma_1} < 2|v(\nu)|^2, \\ &\int_{\Gamma_1}\int_{\Gamma_3} = \int_{\Gamma_3}\int_{\Gamma_1} < 2|v(\nu)|^2, \quad \int_{\Gamma_2}\int_{\Gamma_2} = 0, \\ &\int_{\Gamma_2}\int_{\Gamma_3} = \int_{\Gamma_3}\int_{\Gamma_2} \leq c|v(\nu)|^2, \quad \int_{\Gamma_3}\int_{\Gamma_3} < 2|v(\nu)|^2. \end{aligned}$$

Combining the above inequalities and (4.3.6)—(4.3.9) yields (4.3.5). □

Let u_h be the solution of (4.1.3), $\{u_i^n\}_{i=1}^N$ the sequence resulted from Algorithm 4.2.1. For $i = 1,2,\cdots,N$, let $u_i \in V_h^i$, such that

$$u_i(x) = \begin{cases} u_h(x), & \forall \text{ interpolation point } x \in \overline{\Omega}_i \\ 0, & \forall \text{ interpolation point } x \in \overline{\Omega}\backslash\overline{\Omega}_i \end{cases}$$

Then, $\varepsilon_i^n = u_i^n - u_i$ satisfies

$$\text{(I)}\qquad \varepsilon_i^n \in V_h^i,\ \varepsilon_i^n(x) = \varepsilon_j^n(x),\quad \forall \text{ node } x \in \partial\Omega_i \cap \partial\Omega_j$$
$$A_i(\varepsilon_i^n, \theta) = 0,\quad \forall\, \theta \in V_h^{i,0}$$

$$\text{(II)}\qquad A_i(\psi_i^n, \theta) = \sum_{j=1}^{N} A_j(\varepsilon_j^n, Tr_j^{-1}(r_0^i r_0^h \theta)),\quad \forall\, \theta \in V_h^i$$

Lemma 4.3.3. $$\sum_{i=1}^{N} A_i(\varepsilon_i^n, \varepsilon_i^n) \le \sum_{i=1}^{N} A_i(\psi_i^n, \psi_i^n).$$

Proof. By (I) and (4.2.1), we obtain that $Tr_j^{-1}(\sum_{i=1}^{N} r_0^i r_0^h \varepsilon_i^n) - \varepsilon_j^n \in V_h^{j,0}$. Furthermore, with (I), the substitution of θ in (II) with ε_i^n and the Schwarz inequality, it is easy to see that

$$\begin{aligned}\sum_{j=1}^{N} A_j(\varepsilon_j^n, \varepsilon_j^n) &= \sum_{j=1}^{N} A_j(\varepsilon_j^n, Tr_j^{-1}(\sum_{i=1}^{N} r_0^i r_0^h \varepsilon_i^n))\\ &= \sum_{i=1}^{N}\sum_{j=1}^{N} A_j(\varepsilon_j^n, Tr_j^{-1}(r_0^i r_0^h \varepsilon_i^n)) = \sum_{i=1}^{N} A_i(\psi_i^n, \varepsilon_i^n)\\ &\le \left[\sum_{i=1}^{N} A_i(\psi_i^n, \psi_i^n)\right]^{\frac{1}{2}} \left[\sum_{i=1}^{N} A_i(\varepsilon_i^n, \varepsilon_i^n)\right]^{\frac{1}{2}},\end{aligned}$$

which implies that Lemma 4.3.3 holds. □

Lemma 4.3.4. *Denote* $\varphi_i^n = \frac{1}{\rho}(\varepsilon_i^n - \varepsilon_i^{n+1})$. *we have*

$$\sum_{i=1}^{N} A_i(\varepsilon_i^n, \varphi_i^n) = \sum_{i=1}^{N} A_i(\psi_i^n, \psi_i^n).$$

Proof. The substitution of θ in (I) with $\varphi_i^n - Tr_i^{-1}(\sum_{j=1}^{N} r_0^j r_0^h \psi_j^n) \in V_h^{i,0}$ gives

$$\sum_{i=1}^{N} A_i(\varepsilon_i^n, \varphi_i^n) = \sum_{i=1}^{N} A_i(\varepsilon_i^n, \sum_{j=1}^{N} Tr_i^{-1}(r_0^j r_0^h \psi_j^n)).$$

It follows from (II) with $\theta = \psi_i^n$ that

$$\sum_{i=1}^{N} A_i(\psi_i^n, \psi_i^n) = \sum_{i=1}^{N}\sum_{j=1}^{N} A_j(\varepsilon_j^n, Tr_j^{-1}(r_0^i r_0^h \psi_i^n)) = \sum_{j=1}^{N} A_j(\varepsilon_j^n, \sum_{i=1}^{N} Tr_j^{-1}(r_0^i r_0^h \psi_i^n)).$$

With the above two equalities, we come to the conclusion. □

Lemma 4.3.5. *Let Γ_{jk} be an open edge of Ω_i with endpoints ν_j and ν_k. If $w \in V_h^i$ satisfies*

$$\begin{cases} A_i(w,\theta) = 0, & \forall\ \theta \in V_h^{i,0} \\ w(x) = \frac{1}{2}\psi_i^n(x), & \forall\ \textit{node}\ x \in \Gamma_{jk} \\ w(x) = 0, & \forall\ \textit{node}\ x \in \partial\Omega_i \backslash \Gamma_{jk} \end{cases}$$

then

$$A_i(w,w) \le \tau A_i(\psi_i^n, \psi_i^n),$$

where

$$\tau \le c(1 + \ln\frac{H}{h})(1 + H^{-2} + \ln\frac{H}{h}).$$

Proof. Let $\tilde{w}$ and $\tilde{w}_\perp \in V_h^i$ satisfy respectively

$$\begin{cases} A_i(\tilde{w},\theta) = 0, & \forall\ \theta \in V_h^{i,0} \\ \tilde{w}(x) = \frac{1}{2}\psi_i^n(x), & \forall\ \text{node}\ x \in \partial\Omega_i, \text{where}\ x\ \text{is not any vertex of}\ \Omega_i \\ \tilde{w}(x) = 0, & \forall\ \text{vertex}\ x\ \text{of}\ \Omega_i \end{cases}$$

$$\begin{cases} A_i(\tilde{w}_\perp,\theta) = 0, & \forall\ \theta \in V_h^{i,0} \\ \tilde{w}_\perp(x) = 0, & \forall\ \text{node}\ x \in \partial\Omega_i, \text{where}\ x\ \text{is not any vertex of}\ \Omega_i \\ \tilde{w}_\perp(x) = \frac{1}{2}\psi_i^n(x), & \forall\ \text{vertex}\ x\ \text{of}\ \Omega_i \end{cases}$$

By (II), we have

$$\tilde{w} + \tilde{w}_\perp = \frac{1}{2}\psi_i^n \qquad \text{in}\ \Omega_i. \tag{4.3.10}$$

Denote

$$v = I_h\tilde{w}|_{\partial\Omega_i},\quad v_{jk} = \begin{cases} v, & \text{on}\ \Gamma_{jk} \\ 0, & \text{on}\ \partial\Omega_i\backslash\Gamma_{jk} \end{cases}$$

It follows from the proof of Theorem 2.4.3 that

$$\begin{aligned} A_i(w,w) &\le c\|v_{jk}\|^2_{H_{00}^{\frac{1}{2}}(\Gamma_{jk})} \\ &\le c\Big\{\|v\|^2_{\frac{1}{2},\partial\Omega_i} + \int_{\Gamma_{jk}} \big(\frac{(v_{jk}(x))^2}{|x-\nu_j|} + \frac{(v_{jk}(x))^2}{|x-\nu_k|}\big) ds(x)\Big\}. \end{aligned} \tag{4.3.11}$$

From (4.3.2) and (4.3.3), we know that

$$\|I_h\tilde{w}\|^2_{H^1(\Omega_i)} = \|I_h\tilde{w}\|^2_{L^2(\Omega_i)} + |I_h\tilde{w}|^2_{1,\Omega_i,h} \le c\{\|\tilde{w}\|^2_{L^2(\Omega_i)} + |\tilde{w}|^2_{1,\Omega_i,h}\},$$

by which the trace theorem and the definition of $A(\cdot,\cdot)$ give

$$\|v\|^2_{\frac{1}{2},\partial\Omega_i} = \|I_h\tilde{w}\|^2_{\frac{1}{2},\partial\Omega_i} \le cA_i(\tilde{w},\tilde{w}) \le c\{A_i(\tilde{w}+\tilde{w}_\perp, \tilde{w}+\tilde{w}_\perp) + A_i(\tilde{w}_\perp, \tilde{w}_\perp)\}.$$

Using (4.3.10), Lemma 4.3.2 and the triangle inequality, we obtain

$$\begin{aligned} \|v\|^2_{\frac{1}{2},\partial\Omega_i} &\le c\{A_i(\psi_i^n,\psi_i^n) + \max_{\text{vertex } x \text{ of } \Omega_i} (\psi_i^n(x))^2\} \\ &\le c\{A_i(\psi_i^n,\psi_i^n) + \|\psi_i^n\|^2_{L^\infty(\Omega_i)}\}. \end{aligned} \tag{4.3.12}$$

It follows from Lemma 2.4.5 that

$$\begin{aligned} &\int_{\Gamma_{jk}} \left(\frac{(v_{jk}(x))^2}{|x-\nu_j|} + \frac{(v_{jk}(x))^2}{|x-\nu_k|} \right) ds(x) \\ &\le c(1+\ln\frac{H}{h})\|v_{jk}\|^2_{L^\infty(\Gamma_{jk})} = c(1+\ln\frac{H}{h}) \max_{\text{node } x\in\Gamma_{jk}} (v_{jk}(x))^2 \\ &\le c(1+\ln\frac{H}{h})\|\psi_i^n\|^2_{L^\infty(\Gamma_{jk})} \le c(1+\ln\frac{H}{h})\|\psi_i^n\|^2_{L^\infty(\Omega_i)} \end{aligned} \tag{4.3.13}$$

Now, Theorem 2.4.2 and the substitution of (4.3.12) and (4.3.13) into (4.3.11) imply Lemma 4.3.5 holds. □

Lemma 4.3.6. *Let Γ_{ij} be an open edge with endpoints ν_i and ν_j. Suppose $\overline{\Gamma}_{ij} = \partial\Omega_k \cap \partial\Omega_l$ and $w \in V_h^k$ satisfies*

$$\begin{cases} A_k(w,\theta) = 0, & \forall\, \theta \in V_h^{k,0} \\ w(x) = \frac{1}{2}\psi_l^n(x), & \forall \text{ node } x \in \Gamma_{ij} \\ w(x) = 0, & \forall \text{ node } x \in \partial\Omega_k \backslash \Gamma_{ij} \end{cases}$$

then

$$A_k(w,w) \le \tau A_l(\psi_l^n,\psi_l^n),$$

where τ is the same as that in Lemma 4.3.5.

Proof. Let $\tilde{w} \in V_h^l$ satisfy

$$\begin{cases} A_l(\tilde{w},\theta) = 0, & \forall\, \theta \in V_h^{l,0} \\ \tilde{w}(x) = \frac{1}{2}\psi_l^n(x), & \forall \text{ node } x \in \Gamma_{ij} \\ \tilde{w}(x) = 0, & \forall \text{ node } x \in \partial\Omega_l \backslash \Gamma_{ij} \end{cases}$$

It follows from Theorem 2.4.3 and Lemma 4.3.5 that

$$A_k(w,w) \le cA_l(\tilde{w},\tilde{w}) \le \tau A_l(\psi_l^n,\psi_l^n).$$

Therefore, the lemma holds. □

With the above lemmas, we state and prove the main result.

Theorem 4.3.7. *If $0 < \rho < \frac{2}{\tau}$, then the error sequence $\{\varepsilon_i^n\}_{i=1}^N$ of Algorithm 4.2.1 satisfies*

$$\sum_{i=1}^N A_i(\varepsilon_i^{n+1},\varepsilon_i^{n+1}) \le \kappa(\rho) \sum_{i=1}^N A_i(\varepsilon_i^n,\varepsilon_i^n), \tag{4.3.14}$$

where $\kappa(\rho) = 1-2\rho+\tau\rho^2$, *and* τ *is the same as that in Lemma 4.3.5. Furthermore, we have*

$$0 < \kappa(\rho) < 1, \qquad \text{if } 0 < \rho < \frac{2}{\tau},$$

and

$$\kappa(\rho_{opt}) = 1 - \tfrac{1}{\tau} = \min_{0<\rho<\frac{2}{\tau}} \kappa(\rho) \ , \quad \textit{where } \rho_{opt} = \tfrac{1}{\tau}.$$

Proof. For convenience, we introduce the following notations:

$$\xi_i = \{k : \ \Omega_k \in \Omega_H, \ \text{meas}(\Omega_k \cap \Omega_i) > 0\},$$

$$\eta_j = \{k : \ \Omega_k \in \Omega_H, \ \nu_j \text{ is the vertex of } \Omega_k\}.$$

Let n_j be the number of the elements of η_j. Suppose $\nu_j \in \overline{\Omega}_i$, for $k \in \eta_j$, let $\hat{w}_i^k \in V_h^i$, s.t.

$$\begin{cases} A_i(\hat{w}_i^k, \ \theta) = 0, & \forall\, \theta \in V_h^{i,0} \\ \hat{w}_i^k(\nu_j) = \frac{1}{n_j}\psi_k^n(\nu_j) & \\ \hat{w}_i^k(x) = 0, & \forall \text{ node } x \in \partial\Omega_i, \ x \neq \nu_j \end{cases}$$

By Lemma 4.3.2 and Theorem 2.4.2, we have

$$\begin{aligned} A_i(\hat{w}_i^k, \ \hat{w}_i^k) \ &\leq c|\psi_k^n(\nu_j)|^2 \\ &\leq c \max(H^{-2}, \ 1 + ln\tfrac{H}{h}) A_k(\psi_k^n, \ \psi_k^n) \end{aligned}$$

For $k \in \xi_i$, let $w_i^k \in V_h^i$, such that

$$\begin{cases} A_i(w_i^k, \ \theta) = 0, & \forall\, \theta \in V_h^{i,0} \\ w_i^k(x) = \frac{1}{2}\psi_k^n(x), & \forall \text{ node } x \in (\partial\Omega_i \cap \partial\Omega_k)\backslash\{\nu_j\} \\ w_i^k(x) = 0, & \forall \text{ node } x \in \overline{\partial\Omega_i(\backslash\partial\Omega_i \cap \partial\Omega_k)} \end{cases}$$

It follows from Lemma 4.3.5 and Lemma 4.3.6 that

$$A_i(w_i^k, \ w_i^k) \leq \tau A_k(\psi_k^n, \ \psi_k^n).$$

Furthermore, it is easy to see that

$$\varphi_i^n = \frac{1}{\rho}(\varepsilon_i^n - \varepsilon_i^{n+1}) = \sum_{\nu_j \in \overline{\Omega}_i} \sum_{k \in \eta_j} \hat{w}_i^k + \sum_{k \in \xi_i} w_i^k. \tag{4.3.15}$$

So, with the triangle inequality, the above two inequalities, and the quasi-uniformness of Ω_H, we obtain

$$\begin{aligned} A(\varphi_i^n, \varphi_i^n) \qquad &\leq c\Big\{ \sum_{\nu_j \in \overline{\Omega}_i} \sum_{k \in \eta_j} A_i(\hat{w}_i^k, \ \hat{w}_i^k) + \sum_{k \in \xi_i} A_i(w_i^k, \ w_i^k)\Big\} \\ &\leq \tau\Big\{ \sum_{\nu_j \in \overline{\Omega}_i} \sum_{k \in \eta_j} A_k(\psi_k^n, \ \psi_k^n) + \sum_{k \in \xi_i} A_k(\psi_k^n, \ \psi_k^n)\Big\}, \\ \sum_{i=1}^N A_i(\varphi_i^n, \varphi_i^n) \ &\leq \tau \sum_{i=1}^N A_i(\psi_i^n, \ \psi_i^n). \end{aligned}$$

Hence, by Lemma 4.3.4 and Lemma 4.3.3, we see that for $0 < \rho < \frac{2}{\tau}$

$$\begin{aligned}\sum_{i=1}^{N} A_i(\varepsilon_i^{n+1}, \varepsilon_i^{n+1}) &= \sum_{i=1}^{N} A_i(\varepsilon_i^n, \varepsilon_i^n) - 2\rho \sum_{i=1}^{N} A_i(\varepsilon_i^n, \varphi_i^n) + \rho^2 \sum_{i=1}^{N} A_i(\varphi_i^n, \varphi_i^n) \\ &\leq \sum_{i=1}^{N} A_i(\varepsilon_i^n, \varepsilon_i^n) - 2\rho \sum_{i=1}^{N} A_i(\psi_i^n, \psi_i^n) + \rho^2 \tau \sum_{i=1}^{N} A_i(\psi_i^n, \psi_i^n) \\ &\leq (1 - 2\rho + \tau\rho^2) \sum_{i=1}^{N} A_i(\varepsilon_i^n, \varepsilon_i^n),\end{aligned}$$

which completes the proof of (4.3.14), and hence the theorem. □

4.4 Matrix analysis

Sect.4.3 gives the convergence analysis of Algorithm 4.2.1 by the reduction of error energies. Now, we analyse Algorithm 4.2.1 from the algebraic viewpoint. (4.1.3) can be written as

$$KU = \begin{bmatrix} K_{II} & K_{I\Gamma} \\ K_{I\Gamma}^T & K_{\Gamma\Gamma} \end{bmatrix} \begin{bmatrix} U_I \\ U_\Gamma \end{bmatrix} = \begin{bmatrix} F_I \\ F_\Gamma \end{bmatrix}, \tag{4.4.1}$$

where

$$\begin{aligned} &K = (A(\phi_i, \phi_j)), && F_I = ((f, \phi_i^I)), && F_\Gamma = ((f, \phi_i^\Gamma)), \\ &K_{II} = (A(\phi_i^I, \phi_j^I)), && K_{I\Gamma} = (A(\phi_i^I, \phi_j^\Gamma)), && K_{\Gamma\Gamma} = (A(\phi_i^\Gamma, \phi_j^\Gamma)), \end{aligned}$$

$\{\phi_i\}$ represents the set of the basis functions of V_h^0, ϕ_i^I and ϕ_j^Γ are a basis function of V_h^0 in $\bigcup_{k=1}^{N} \Omega_k$ and that on Γ, respectively. By the construction of the nonconforming elements, U_Γ is the vector of the node values of u_h on Γ. With the block Gauss elimination, we obtain

$$SU_\Gamma = F_\Gamma - K_{I\Gamma}^T K_{II}^{-1} F_I = \tilde{F}_\Gamma, \tag{4.4.2}$$

where $S = K_{\Gamma\Gamma} - K_{I\Gamma}^T K_{II}^{-1} K_{I\Gamma}$ is the capacitance matrix of the stiff matrix K with regard to Γ. Of course, once the solution U_Γ of the capacitance equation (4.4.2) is known, the solution U of (4.4.1) can be obtained by solving in parallel the discrete problems on subdomains (cf. step 2 of Algorithm 4.2.1).

Let $\{\theta_j\}$ be the set of the basis function of V_h^i. θ_j^I and θ_k^Γ represent a basis function of V_h^i in Ω_i and that on $\partial\Omega_i \backslash \partial\Omega$, respectively. Denote

$$K_{ii} = (A_i(\theta_j^I, \theta_k^I)), \quad K_{i\Gamma_i} = (A_i(\theta_j^I, \theta_k^\Gamma)), \quad K_{\Gamma_i\Gamma_i} = (A_i(\theta_j^\Gamma, \theta_k^\Gamma)),$$

$$S_i = K_{\Gamma_i\Gamma_i} - K_{i\Gamma_i}^T K_{ii}^{-1} K_{i\Gamma_i}, \quad R_i = ((r_0^i)_{jk}, k \in I_i) \in \Re^{m \times m_i},$$

where r_0^i, I_i and m_i are the same as those in Sect.4.2. Correspondingly, S_i is called the capacitance matrix of K in Ω_i concerning $\partial\Omega_i \backslash \partial\Omega$. It is easy to see that Algorithm 4.2.1 is essentially equivalent to the following iterative method for (4.4.2):

$$\lambda^{n+1} = \lambda^n - \rho \sum_{i=1}^{N} R_i S_i^{-1} R_i^T S \lambda^n + \rho \sum_{i=1}^{N} R_i S_i^{-1} R_i^T \tilde{F}_\Gamma. \tag{4.4.3}$$

Obviously, (4.4.3) can be viewed as the simple iterative method for (4.4.2) preconditioned by

$$Q = (\rho \sum_{i=1}^{N} R_i S_i^{-1} R_i^T)^{-1}. \tag{4.4.4}$$

For the simple iterative method converges much more slowly than the conjugate gradient method (CG) to solve the same symmetric positive definite algebraic equation, we consider applying CG to (4.4.2) with (4.4.4) as preconditioner. In fact, Q is an efficient preconditioner of the capacitance matrix S, because it can be inversed easily in parallel and the condition number of $Q^{-1}S$ is bounded by $O\Big((1+\ln \frac{H}{h})(1+H^{-2}+\ln \frac{H}{h})\Big)$, which is implied by the following theorem.

Theorem 4.4.1. *Let S and Q be those in (4.4.2) and (4.4.4); then for arbitrary $\lambda \in \Re^m \backslash \{0\}$,*

$$\rho \leq \frac{(SQ^{-1}S\lambda, \lambda)}{(S\lambda, \lambda)} \leq c\rho(1 + \ln \frac{H}{h})(1 + H^{-2} + \ln \frac{H}{h}). \tag{4.4.5}$$

Proof. Let ε_i, ψ_i be generated by Algorithm 4.2.1 with $f = 0$ and $\lambda^n = \lambda$. A more iteration gives $\bar{\varepsilon}_i, \bar{\psi}_i$. Denote $\varphi_i = \frac{1}{\rho}(\varepsilon_i - \bar{\varepsilon}_i)$. It is easy to see that

$$(S\lambda, \lambda) = \sum_{i=1}^{N} A_i(\varepsilon_i, \varepsilon_i), \quad (SQ^{-1}S\lambda, \lambda) = \rho \sum_{i=1}^{N} A_i(\varepsilon_i, \varphi_i).$$

Hence, by Lemma 4.3.4 and Lemma 4.3.3, we get the left part of (4.4.5).

On the other hand, it follows from Lemma 4.3.4 and (4.3.16) that

$$\begin{aligned} \sum_{i=1}^{N} A_i(\psi_i, \psi_i) &= \sum_{i=1}^{N} A_i(\varepsilon_i, \varphi_i) \\ &\leq [\sum_{i=1}^{N} A_i(\varepsilon_i, \varepsilon_i)]^{\frac{1}{2}} [\sum_{i=1}^{N} A_i(\varphi_i, \varphi_i)]^{\frac{1}{2}} \\ &\leq [\sum_{i=1}^{N} A_i(\varepsilon_i, \varepsilon_i)]^{\frac{1}{2}} [\tau \sum_{i=1}^{N} A_i(\psi_i, \psi_i)]^{\frac{1}{2}}, \end{aligned}$$

which implies the right part of (4.4.5) holds, and hence the theorem. □

Remark 4.4.1. When $N = 2$, Algorithm 4.2.1 degenerates to be Algorithm 3.4.2.

Remark 4.4.2. It follows from the properties of Q defined by (4.4.4) that

$$\begin{bmatrix} K_{II} & K_{I\Gamma} \\ K_{I\Gamma}^T & Q + K_{I\Gamma}^T K_{II}^{-1} K_{I\Gamma} \end{bmatrix}$$

is an efficient preconditioner of the stiff matrix in (4.4.1).

Chapter 5

Preconditioners with Multi–subdomains for First Kind Nonconforming Elements by Substructuring

5.1 Introduction

Nonoverlap domain decomposition methods (substructuring methods), with a long history from the structural analysis community[75], are methods where the global domain is partitioned into disjoint smaller subdomains. More recently, the construction of preconditioners by substructuring has been investigated for the solving of the conforming finite element discretizations of elliptic problems (see [12–15] and other references in this book)

$$u \in H_0^1(\Omega): \ a(u,v) = (f,v), \ \forall\, v \in H_0^1(\Omega) \tag{5.1.1}$$

where $a(u,v) = \int_\Omega \left[\sum_{i,j=1}^{2} a_{ij}(x) \frac{\partial u}{\partial x_i} \frac{\partial v}{\partial x_j} \right]$, $(f,v) = \int_\Omega fv$, $f \in H^{-1}(\Omega)$, $\Omega \subset \Re^2$ is a bounded polygonal open domain, $a_{ij}(x)$ $(i,j = 1,2)$ are piecewise smooth and bounded functions in Ω and (a_{ij}) is a symmetric, uniformly positive definite matrix in Ω.

The aim of this chapter is to construct preconditioners for the discretization of (5.1.1) via the nonconforming finite elements of the first kind. Let $\Omega_h = \{e\}$ be a quasi–uniform mesh of Ω with h as its mesh parameter. Let V_h be a nonconforming finite element space of the first kind defined in Chapter 1. The nonconforming finite element discretization of (5.1.1) is

$$u_h \in V_h^0: \ A(u_h,v) = (f,v), \ \ \forall\, v \in V_h^0 \tag{5.1.2}$$

where $\qquad V_h^0 = \{v \in V_h: \ v(x) = 0, \ \forall \text{ interpolation point } x \in \partial\Omega\},$

$$A(w,v) = \sum_{e \subset \Omega} \int_e \left[\sum_{i,j=1}^{2} a_{ij} \frac{\partial w}{\partial x_i} \frac{\partial v}{\partial x_j} \right].$$

Suppose that $\{\Omega_k\}_{k=1}^N \triangleq \Omega_H$ is a nonoverlap subdivision of Ω, which is, in fact, a quasi–uniform mesh of Ω with H as its mesh parameter. each Ω_i is either a quadrilateral or a triangle, referred to as the subdomain. Besides, the common boundary point of more than two subdomains that lies inside Ω is called the internal cross point. Assume that Ω_h can be obtained by refining Ω_H, so that Ω_H and Ω_h

form a two–level triangulation[7,16,20,32] on Ω. Generally, $H \gg h$. Therefore, Ω_H and Ω_h are called the coarse mesh and the fine mesh respectively. If all the subdomains are colored by two colors, red and black, such that no pair of subdomains which have a common edge have the same color, then such a subdivision of Ω is called the red–black subdivision. We note that the red–black subdivision is of course not always possible. In the conforming discrete cases, some scholars have developed many preconditioners which are easily applied in parallel and can decrease the condition number of the discrete system from $O(h^{-2})$ to $O\Big((1+\ln\frac{H}{h})^k\Big)$, $(k=2,3)$. When handling the nonconforming discrete problem (5.1.2), we encounter two difficulties: how to cope with the internal cross points and the lack of global continuity of functions in V_h.

To overcome the former, a coarse mesh problem must be solved, which is employed the idea of Bramble[12], in order to separate the internal cross points from other mesh nodes. We note that some essential estimates have been established in Chapter 2 where the conforming interpolation operator I_h is introduced to act as a bridge between the nonconforming element space V_h and the conforming element space where many fundamental inequalities have already been constructed. Our preconditioners are constructed for (5.1.2) in the red–black subdivisional case and in the general case. The preconditioned algorithms are high parallel and the condition number of the preconditioned system is either $O\Big((1+\ln\frac{H}{h})^2\Big)$ or $O\Big((1+\ln\frac{H}{h})^3\Big)$. Thus, they are very efficient.

In the remainder of the present chapter, the notations and lemmas are given in Sect.5.2. A series of preconditioners are constructed and analysed in Sect.5.3.

5.2 Preliminary

Let $\{\xi_j\}_{j=1}^m$ be the set of the vertices of Ω_k, $k=1,2,\cdots,N$ (ordered in some way). *The conforming interpolation operator* $I_h : V_h \to C(\Omega)$ is the same as that in Chapter 4, which is defined by $\forall\, v \in V_h$, $I_h v \in C(\Omega)$ such that

$$(I_h v)(x) = v(x),\ \forall \text{ vertex } x \text{ of } e,\ I_h v|_e \text{ is linear (bilinear)},\ \forall\, e \in \Omega_h.$$

Let $\Gamma_{ij} \not\subset \partial\Omega$ be the open edge with ξ_i and ξ_j as its endpoints which is a boundary edge of some subdomain Ω_k. And, we introduce the following notations:

$$V_h^k = \{v \in V_h^0 :\ v(x) = 0,\ \forall \text{ interpolation point } x \in \Omega\backslash\overline{\Omega}_k\},$$

$$V_h^{k,0} = \{v \in V_h^0 :\ v(x) = 0,\ \forall \text{ interpolation point } x \in \Omega\backslash\Omega_k\},$$

$$\beta_k = \{(i,j) :\ \Gamma_{ij} \subset \partial\Omega_k,\ 1 \le i,j \le m\},$$

$$A_k(w,v) = \sum_{e\subset\Omega_k} \int_e \Big[\sum_{i,j=1}^{2} a_{ij}\frac{\partial w}{\partial x_i}\frac{\partial v}{\partial x_j}\Big].$$

If $v \in V_h$ satisfies

$$A_k(v,w) = 0,\ \forall\, w \in V_h^{k,0},$$

then v is called *a discrete harmonic function* on Ω_k. For each $v \in V_h$, $\overline{v}_{ij}$ denotes the piecewise linear continuous function on $\bigcup_{k=1}^{N} \partial\Omega_k$, which satisfies

$$\overline{v}_{ij}(x) = \begin{cases} v(x), & \forall \text{ interpolation point } x \in \overline{\Gamma_{ij}} \\ 0, & \forall \text{ interpolation point } x \in \bigcup_{k=1}^{N} \partial\Omega_k \backslash \overline{\Gamma_{ij}} \end{cases} \tag{5.2.1}$$

And let $\hat{v}_{ij} \in V_h$ be the discrete harmonic function on Ω_k, $k = 1, 2, \cdots, N$, s.t.

$$\hat{v}_{ij}(x) = \overline{v}_{ij}(x), \ \forall \text{ interpolation point } x \in \bigcup_{k=1}^{N} \partial\Omega_k \tag{5.2.2}$$

Besides, v_{ij} denotes the vector of the values of v at the interpolation points on Γ_{ij}. Let $I_H : V_h \to C(\Omega)$ be *the coarse mesh interpolation operator*, s.t.

$$\forall \ v \in V_h, \ I_H v \in C(\Omega) : \begin{cases} (I_H v)(\xi_j) = v(\xi_j), \ j = 1, 2, \cdots, m \\ I_H v|_{\Omega_k} \text{ is linear (bilinear)}, \ k = 1, 2, \cdots, N \end{cases}$$

Let $S_0^h(\Gamma_{ij}) \subset H_0^1(\Gamma_{ij})$ be the piecewise linear continuous finite element space. We note that the operator $l_{\Gamma_{ij}}$ defined on $S_0^h(\Gamma_{ij})$ by

$$\langle l_{\Gamma_{ij}} v, w \rangle_{\Gamma_{ij}} = \int_{\Gamma_{ij}} \frac{dv}{ds}\frac{dw}{ds} ds, \ \forall \ w \in S_0^h(\Gamma_{ij}) \tag{5.2.3}$$

is a finite–dimensional approximation to the Laplace operator $-\frac{d^2}{ds^2}$. Here, s is the arc length along Γ_{ij}. Let $l_{\Gamma_{ij}}^{\frac{1}{2}}$ be the positive square root of the symmetric, positive definite matrix $l_{\Gamma_{ij}}$.

Lemma 5.2.1. *1) If $v \in V_h$ is a discrete harmonic function on Ω_k, then*

$$A_k(v, v) \le c\|I_h v\|^2_{\frac{1}{2}, \partial\Omega_k}.$$

2) If $v \in V_h$ satisfies

$$\int_{\Omega_k} v = 0 \ \textit{or} \ v(x) = 0, \ \forall \ \textit{interpolation point} \ x \in \Gamma,$$

where $\Gamma \subset \partial\Omega_k$ has at least two interpolation points, then

$$A_k(v, v) \ge c\|I_h v\|^2_{\frac{1}{2}, \partial\Omega_k}.$$

Proof. 1) has been shown by Theorem 2.4.3. Thus we need only prove 2). It follows from Lemma 2.4.1 1) or Lemma 2.4.1 3) that

$$\|v\|^2_{L^2(\Omega_k)} \le cH^2|v|^2_{1,\Omega_k,h}.$$

Therefore, by Lemma 4.3.1 and the trace theorem, we obtain

$$\begin{aligned}\|I_h v\|^2_{\frac{1}{2},\partial\Omega_k} &\le c\|I_h v\|^2_{H^1(\Omega_k)} = c\Big\{\|I_h v\|^2_{L^2(\Omega_k)} + |I_h v|^2_{1,\Omega_k,h}\Big\} \\ &\le c\Big\{\|v\|^2_{L^2(\Omega_k)} + |v|^2_{1,\Omega_k,h}\Big\} \le c|v|^2_{1,\Omega_k,h} \le cA_k(v,v).\end{aligned}$$

Here, we assume that $H \le c_0$, where c_0 is some positive constant, e.g., $c_0 = 1$, hence c is independent of H. By now, Lemma 5.2.1 follows. □

Lemma 5.2.2. *Let $v \in V_h$ be a discrete harmonic function on Ω_k.*
1) If v vanishes at the vertices of Ω_k, then

$$A_k(v,v) \le c \sum_{(i,j)\in\beta_k} \langle l^{\frac{1}{2}}_{\Gamma_{ij}} \overline{v}_{ij}, \overline{v}_{ij}\rangle_{\Gamma_{ij}}.$$

2) If v is linear on Γ_{ij}, $\forall\ (i,j) \in \beta_k$, then

$$A_k(v,v) \le c \sum_{(i,j)\in\beta_k} \Big(v(\xi_i) - v(\xi_j)\Big)^2.$$

Proof. Let's show 1) first. Note that $I_h v = \sum_{(i,j)\in\beta_k} \overline{v}_{ij}$ on $\partial\Omega_k$. Therefore, by Lemma 5.2.1 1) and the triangle inequality, we have

$$\begin{aligned}A_k(v,v) \le c\|I_h v\|^2_{\frac{1}{2},\partial\Omega_k} &\le c \sum_{(i,j)\in\beta_k} \|\overline{v}_{ij}\|^2_{\frac{1}{2},\partial\Omega_k} \\ &\le c \sum_{(i,j)\in\beta_k} \|\overline{v}_{ij}\|^2_{H^{\frac{1}{2}}_{00}(\Gamma_{ij})} \le c \sum_{(i,j)\in\beta_k} \langle l^{\frac{1}{2}}_{\Gamma_{ij}} \overline{v}_{ij}, \overline{v}_{ij}\rangle_{\Gamma_{ij}},\end{aligned}$$

where the last two inequalities hold because of the equivalence[12,60] of $\|w\|^2_{\frac{1}{2},\partial\Omega_k}$, $\|w\|^2_{H^{\frac{1}{2}}_{00}(\Gamma_{ij})}$ and $\langle l^{\frac{1}{2}}_{\Gamma_{ij}} w, w\rangle_{\Gamma_{ij}}$ for $w \in C(\partial\Omega_k)$ with $w = 0$ on $\partial\Omega_k \backslash \Gamma_{ij}$.

Next we prove 2). Let $\tilde{v}$ be the linear (bilinear) function on Ω_k, s.t. $\tilde{v} = v$ on $\partial\Omega_k$. Denote $\beta = \frac{1}{\text{meas}(\Omega_k)} \int_{\Omega_k} \tilde{v}$. Then $v - \beta$ is also a discrete harmonic function on Ω_k. Hence, it follows from Lemma 5.2.1 1) that

$$A_k(v,v) = A_k(v-\beta, v-\beta) \le c\|I_h(v-\beta)\|^2_{\frac{1}{2},\partial\Omega_k} = c\|v-\beta\|^2_{\frac{1}{2},\partial\Omega_k}.$$

By the trace theorem and Lemma 1.2.2, we obtain

$$\|v-\beta\|^2_{\frac{1}{2},\partial\Omega_k} = \|\tilde{v}-\beta\|^2_{\frac{1}{2},\partial\Omega_k} \le c\|\tilde{v}-\beta\|^2_{H^1(\Omega_k)} \le c|\tilde{v}-\beta|^2_{H^1(\Omega_k)} = c|\tilde{v}|^2_{H^1(\Omega_k)}.$$

A simple calculation yields

$$|\tilde{v}|^2_{H^1(\Omega_k)} \le c \sum_{(i,j)\in\beta_k} \Big(v(\xi_i) - v(\xi_j)\Big)^2.$$

With the above inequalities, we end the proof of Lemma 5.2.2 2). □

Lemma 5.2.3. *1) If $v \in V_h$ satisfies $v(p) = 0$ for some point $p \in \overline{\Omega}_k$, then*

$$\max_{e \subset \Omega_k} \|v\|^2_{L^\infty(e)} \le c\Big(1 + \ln\frac{H}{h}\Big) A_k(v, v).$$

2)
$$\sum_{(i,j)\in\beta_k} \Big(v(\xi_i) - v(\xi_j)\Big)^2 \le c\Big(1 + \ln\frac{H}{h}\Big) A_k(v, v), \quad \forall\, v \in V_h$$

Proof. Obviously, 2) is the direct corollary of 1). It suffices to show 1). Denote $\beta = \frac{1}{\text{meas}(\Omega_k)} \int_{\Omega_k} v$. Lemma 2.4.1 1) indicates that

$$H^{-2}\|v - \beta\|^2_{L^2(\Omega_k)} \le c|v - \beta|^2_{1,\Omega_k,h} = c|v|^2_{1,\Omega_k,h} \le cA_k(v, v).$$

Furthermore, by Lemma 2.4.2, we obtain

$$\max_{e \subset \Omega_k} \|v - \beta\|^2_{L^\infty(e)} \le c\Big(1 + \ln\frac{H}{h}\Big) A_k(v, v),$$

$$|\beta|^2 = |\beta - v(p)|^2 \le \max_{e \subset \Omega_k} \|v - \beta\|^2_{L^\infty(e)}.$$

Hence, applying the triangle inequality leads to Lemma 5.2.3 1). □

Lemma 5.2.4. *Let $v \in V_h$ s.t. $v(x) = 0$, $\forall\, x \in \{\xi_j\} \bigcap \partial\Omega_k$. If v_L is a linear (bilinear) function on Ω_k or v_L is discrete harmonic on Ω_k and linear on Γ_{ij}, $\forall\, (i,j) \in \beta_k$, then*

$$\langle l^{\frac{1}{2}}_{\Gamma_{ij}} \overline{v}_{ij}, \overline{v}_{ij} \rangle_{\Gamma_{ij}} \le c\Big(1 + \ln\frac{H}{h}\Big)^2 A_k(v + v_L, v + v_L), \ \forall\, (i,j) \in \beta_k \tag{5.2.4}$$

Proof. Let's show (5.2.4) in the case that $v_L = 0$. By the definition of the norm of $H^{\frac{1}{2}}_{00}(\Gamma_{ij})$ (see, e.g., [12,60]), we have

$$\begin{aligned} \langle l^{\frac{1}{2}}_{\Gamma_{ij}} \overline{v}_{ij}, \overline{v}_{ij} \rangle_{\Gamma_{ij}} &\le c\|\overline{v}_{ij}\|^2_{H^{\frac{1}{2}}_{00}(\Gamma_{ij})} \\ &\le c\Big\{\|I_h v\|^2_{\frac{1}{2},\partial\Omega_k} + \int_{\Gamma_{ij}} \Big[\frac{(\overline{v}_{ij}(x))^2}{|x - \xi_i|} + \frac{(\overline{v}_{ij}(x))^2}{|x - \xi_j|}\Big] ds(x)\Big\}. \end{aligned} \tag{5.2.5}$$

Denote $\beta = \frac{1}{\text{meas}(\Omega_k)} \int_{\Omega_k} v$. Applying Lemma 2.4.2, Lemma 5.2.1 2), and Lemma 2.4.1 1) leads to

$$\begin{aligned} \|I_h v\|^2_{\frac{1}{2},\partial\Omega_k} &\le 2\{|\beta|^2 + \|I_h v - \beta\|^2_{\frac{1}{2},\partial\Omega_k}\} \\ &\le 2\{\max_{e \subset \Omega_k} \|v - \beta\|^2_{L^\infty(e)} + \|I_h(v - \beta)\|^2_{\frac{1}{2},\partial\Omega_k}\} \\ &\le c\{H^{-2}\|v - \beta\|^2_{L^2(\Omega_k)} + (1 + \ln\tfrac{H}{h})|v - \beta|^2_{1,\Omega_k,h} + A_k(v - \beta, v - \beta)\} \\ &\le c(1 + \ln\tfrac{H}{h}) A_k(v, v). \end{aligned}$$

On the other hand, with Lemma 2.4.5 and Lemma 5.2.3 1), we obtain that

$$\begin{aligned}\int_{\Gamma_{ij}} \frac{(\overline{v}_{ij}(x))^2}{|x-\xi_i|} ds(x) &\le c\Big(1+\ln\tfrac{H}{h}\Big)\|\overline{v}_{ij}\|^2_{L^\infty(\Gamma_{ij})} \\ &\le c\Big(1+\ln\tfrac{H}{h}\Big)\max_{e\subset\Omega_k}\|v\|^2_{L^\infty(e)} \\ &\le c\Big(1+\ln\tfrac{H}{h}\Big)^2 A_k(v,v), \\ \int_{\Gamma_{ij}} \frac{(\overline{v}_{ij}(x))^2}{|x-\xi_j|} ds(x) &\le c\Big(1+\ln\tfrac{H}{h}\Big)^2 A_k(v,v).\end{aligned}$$

Combining the above inequalities with (5.2.5), we see that (5.2.4) is true in the case that $v_L = 0$.

To prove (5.2.4) in the general case, Let $v_\perp \in V_h^k$, s.t.

$$\begin{cases} A_k(v_\perp, w) = 0, & \forall\, w \in V_h^k \text{ such that } w(x) = 0,\ \forall\, x \in \{\xi_j\}\bigcap \partial\Omega_k \\ v_\perp(x) = v_L(x), & \forall\, x \in \{\xi_j\}\bigcap \partial\Omega_k \end{cases}$$

Then
$$(v + v_L - v_\perp)(x) = (v_L - v_\perp)(x) = 0, \quad \forall\, x \in \{\xi_j\}\bigcap \partial\Omega_k$$

$$A_k(v_\perp, v + v_L - v_\perp) = A_k(v_\perp, v_L - v_\perp) = 0. \tag{5.2.6}$$

Hence,
$$A_k(v + v_L - v_\perp, v + v_L - v_\perp) \le A_k(v + v_L, v + v_L).$$

Furthermore, the special case of (5.2.4) proved above gives

$$\begin{aligned}\langle l^{\frac12}_{\Gamma_{ij}}\overline{v}_{ij}, \overline{v}_{ij}\rangle_{\Gamma_{ij}} &\le 2\langle l^{\frac12}_{\Gamma_{ij}}\overline{(v_L - v_\perp)_{ij}}, \overline{(v_L - v_\perp)_{ij}}\rangle_{\Gamma_{ij}} \\ &\quad + 2\langle l^{\frac12}_{\Gamma_{ij}}\overline{(v + v_L - v_\perp)_{ij}}, \overline{(v + v_L - v_\perp)_{ij}}\rangle_{\Gamma_{ij}} \\ &\le 2\langle l^{\frac12}_{\Gamma_{ij}}\overline{(v_L - v_\perp)_{ij}}, \overline{(v_L - v_\perp)_{ij}}\rangle_{\Gamma_{ij}} \\ &\quad + c\Big(1+\ln\tfrac{H}{h}\Big)^2 A_k(v + v_L - v_\perp, v + v_L - v_\perp) \\ &\le 2\langle l^{\frac12}_{\Gamma_{ij}}\overline{(v_L - v_\perp)_{ij}}, \overline{(v_L - v_\perp)_{ij}}\rangle_{\Gamma_{ij}} \\ &\quad + c\Big(1+\ln\tfrac{H}{h}\Big)^2 A_k(v + v_L, v + v_L).\end{aligned}$$

Thus to complete the proof of the lemma we need only show that

$$\langle l^{\frac12}_{\Gamma_{ij}}\overline{(v_L - v_\perp)_{ij}}, \overline{(v_L - v_\perp)_{ij}}\rangle_{\Gamma_{ij}} \le c\Big(1+\ln\frac{H}{h}\Big)^2 A_k(v + v_L, v + v_L). \tag{5.2.7}$$

Since $v_L - v_\perp$ vanishes at the vertices of Ω_k, applying inequality (5.2.5) and the subsequent arguments gives

$$\langle l^{\frac12}_{\Gamma_{ij}}\overline{(v_L - v_\perp)_{ij}}, \overline{(v_L - v_\perp)_{ij}}\rangle_{\Gamma_{ij}} \le c\Big(1+\ln\frac{H}{h}\Big) A_k(v_L - v_\perp, v_L - v_\perp) + \Theta_1 + \Theta_2, \tag{5.2.8}$$

where

$$\Theta_1 = c\int_{\Gamma_{ij}} \frac{|\overline{(v_L - v_\perp)}_{ij}(x)|^2}{|x-\xi_i|} ds(x), \quad \Theta_2 = c\int_{\Gamma_{ij}} \frac{|\overline{(v_L - v_\perp)}_{ij}(x)|^2}{|x-\xi_j|} ds(x).$$

Since $v_\perp$ is orthogonal to $v_L - v_\perp$ in the $A_k(\cdot,\cdot)$–inner product, we have in view of Lemma 5.2.3 2) that

$$\begin{aligned} A_k(v_L - v_\perp, v_L - v_\perp) &\le A_k(v_L, v_L) \\ &\le c \sum_{(i,j)\in\beta_k} \Big(v_L(\xi_i) - v_L(\xi_j)\Big)^2 \\ &\le c \sum_{(i,j)\in\beta_k} \Big((v+v_L)(\xi_i) - (v+v_L)(\xi_j)\Big)^2 \\ &\le c\Big(1+\ln\tfrac{H}{h}\Big) A_k(v+v_L, v+v_L), \end{aligned} \tag{5.2.9}$$

where the second inequality holds by a simple calculation if v_L is linear (bilinear) on Ω_k, or by Lemma 5.2.2 2) if v_L is discrete harmonic on Ω_k and linear on $\Gamma_{ij} \subset \partial\Omega_k$, $\forall\ (i,j) \in \beta_k$.

The triangle inequality leads to

$$\Theta_1 \le \Theta_{11} + \Theta_{12}, \tag{5.2.10}$$

where

$$\Theta_{11} = c\int_{\Gamma_{ij}} \frac{|v_L(x) - v_L(\xi_i)|^2}{|x-\xi_i|} ds(x), \ \Theta_{12} = c\int_{\Gamma_{ij}} \frac{|\overline{(v_\perp)}_{ij}(x) - v_\perp(\xi_i)|^2}{|x-\xi_i|} ds(x).$$

(5.2.6) implies that

$$A_k(v_\perp, v_\perp) \le A_k(v+v_L, v+v_L).$$

Furthermore, it follows from the linearity of v_L on Γ_{ij}, Lemma 5.2.3 and Lemma 2.4.5 that

$$\begin{aligned} \Theta_{11} &\le c\Big(v_L(\xi_j) - v_L(\xi_i)\Big)^2 \\ &= c\Big[(v+v_L)(\xi_j) - (v+v_L)(\xi_i)\Big]^2 \\ &\le c\Big(1+\ln\tfrac{H}{h}\Big) A_k(v+v_L, v+v_L), \\ \Theta_{12} &\le c\Big(1+\ln\tfrac{H}{h}\Big) \|\overline{(v_\perp)}_{ij} - v_\perp(\xi_i)\|^2_{L^\infty(\Gamma_{ij})} \\ &\le c\Big(1+\ln\tfrac{H}{h}\Big) \max_{e\subset\Omega_k} \|v_\perp - v_\perp(\xi_i)\|^2_{L^\infty(e)} \\ &\le c\Big(1+\ln\tfrac{H}{h}\Big)^2 A_k(v_\perp, v_\perp) \\ &\le c\Big(1+\ln\tfrac{H}{h}\Big)^2 A_k(v+v_L, v+v_L). \end{aligned}$$

In view of (5.2.10), we have $\Theta_1 \le c\Big(1+\ln\frac{H}{h}\Big)^2 A_k(v+v_L, v+v_L)$.
By symmetry, we also have $\Theta_2 \le c\Big(1+\ln\frac{H}{h}\Big)^2 A_k(v+v_L, v+v_L)$.

Thus, (5.2.7) follows from (5.2.8) and (5.2.9). □

5.3 Preconditioners

Let $\{\phi_i\}_1^M$ be the set of the bases of V_h^0 and $u_h = \sum_{i=1}^{M} u_i\phi_i$ the solution of (5.1.2). Then, (5.1.2) can be rewritten as

$$\mathcal{A}U = F, \tag{5.3.1}$$

where $U = (u_i) \in \Re^M$, $F = ((f,\phi_i)) \in \Re^M$, and $\mathcal{A} = (A(\phi_i,\phi_j)) \in \Re^{M\times M}$. The condition number of the symmetric, positive definite matrix $\mathcal{A}$ is bounded by ch^{-2} according to the following theorem.

Theorem 5.3.1. *If Ω_h is quasi–uniform, then cond$(\mathcal{A}) = O(h^{-2})$.*

As in the conforming cases, we can show Theorem 5.3.1 by the Rayleigh theorem, the inverse inequality and Lemma 2.4.1 2). For details, refer to [26,41,47].

When h is small enough, $\mathcal{A}$ is rather ill–conditioned. The preconditioned conjugate gradient method (PCG) is preferable to solve (5.1.2). For this purpose, another symmetric, positive definite matrix $\mathcal{B}$, which is called the preconditioner, must be constructed to satisfy
(1) $\mathcal{B}^{-1}b$ is easier to obtain than $\mathcal{A}^{-1}b$ in parallel computing environments;
(2) $\mathcal{B}$ is spectrally close to $\mathcal{A}$ in the sense that *cond*$(\mathcal{B}^{-1}\mathcal{A})$ should not be large.
Correspondingly, they are equivalent to find a bilinear form $B(\cdot,\cdot)$, such that
(1) For a given function g, the solution v of the following problem is easier in parallel to be obtained than that of (5.1.2)

$$v \in V_h^0 : \ B(v,w) = (g,w), \ \forall\ w \in V_h^0 \tag{5.3.2}$$

(2) There exist two constants λ_0 and λ_1; $\frac{\lambda_1}{\lambda_0}$ is as small as possible, s.t.

$$\lambda_0 B(w,w) \le A(w,w) \le \lambda_1 B(w,w), \ \forall\ w \in V_h^0 \tag{5.3.3}$$

$B(\cdot,\cdot)$ is also called the preconditioner for convenience. To this end, we decompose $w \in V_h^0$ into $w = w_p + w_H$, $w_H = w_E + w_V$, s.t.

$$\begin{cases} w_p \in V_h^{1,0} \oplus V_h^{2,0} \oplus \cdots \oplus V_h^{N,0} \\ A(w_p, v) = A(w,v), \ \forall\ v \in V_h^{1,0} \oplus V_h^{2,0} \oplus \cdots \oplus V_h^{N,0} \end{cases} \tag{5.3.4}$$

$$\begin{cases} w_H \in V_h^0 \\ A(w_H, v) = 0, \quad \forall\ v \in V_h^{1,0} \oplus V_h^{2,0} \oplus \cdots \oplus V_h^{N,0} \\ w_H(x) = w(x), \ \ \forall \text{ interpolation point } x \in \bigcup_{k=1}^{N} \partial\Omega_k \end{cases} \tag{5.3.5}$$

$$\begin{cases} w_V \in V_h^0 \\ A(w_V, v) = 0, \qquad \forall\, v \in V_h^{1,0} \oplus V_h^{2,0} \oplus \cdots \oplus V_h^{N,0} \\ w_V(\xi_j) = w_H(\xi_j), \quad j = 1, 2, \cdots, m \\ w_V|_{\Gamma_{ij}} \text{ is linear}, \qquad \forall\, \Gamma_{ij} \end{cases} \tag{5.3.6}$$

$$\begin{cases} w_E \in V_h^0 \\ A(w_E, v) = 0, \qquad \forall\, v \in V_h^{1,0} \oplus V_h^{2,0} \oplus \cdots \oplus V_h^{N,0} \\ w_E(x) = w_H(x) - w_V(x), \quad \forall \text{ interpolation point } x \in \bigcup_{k=1}^{N} \partial\Omega_k \end{cases} \tag{5.3.7}$$

Obviously, w_H, w_E, w_V are discrete harmonic on Ω_k, $k = 1, 2, \cdots, N$, respectively. In addition, we have

$$A(w, w) = A(w_p, w_p) + A(w_H, w_H). \tag{5.3.8}$$

Note that w_p can be obtained by solving previously independent subproblems (5.3.4). Thus, the key to constructing the preconditioner $B(\cdot,\cdot)$ is to find some proper form to replace $A(w_H, w_H) = \sum_{k=1}^{N} A_k(w_H, w_H)$ in (5.3.8).

Theorem 5.3.2. *The bilinear form*

$$B_1(v, w) = A(v_p, w_p) + \sum_{\Gamma_{ij}} \langle l_{\Gamma_{ij}}^{\frac{1}{2}} \overline{(v_E)}_{ij}, \overline{(w_E)}_{ij} \rangle_{\Gamma_{ij}} + A(I_H v, I_H w), \tag{5.3.9}$$

satisfies

$$c\Big(1 + \ln \frac{H}{h}\Big)^{-2} B_1(w, w) \le A(w, w) \le C B_1(w, w), \ \forall\, w \in V_h^0 \tag{5.3.10}$$

Proof. By the proof of Lemma 5.2.2 2), we see that

$$A_k(w_V, w_V) \le c A_k(I_H w, I_H w).$$

Furthermore, Lemma 5.2.2 1) leads to

$$\begin{aligned} A_k(w_H, w_H) &\le 2A_k(w_E, w_E) + 2A_k(w_V, w_V) \\ &\le c\Big\{ \sum_{(i,j)\in\beta_k} \langle l_{\Gamma_{ij}}^{\frac{1}{2}} \overline{(w_E)}_{ij}, \overline{(w_E)}_{ij} \rangle_{\Gamma_{ij}} + A_k(I_H w, I_H w) \Big\}, \end{aligned}$$

which implies that the right hand side of (5.3.10) holds.

On the other hand, it follows from Lemma 5.2.3 2) and Lemma 5.2.4 that

$$A_k(I_H w, I_H w) \le c \sum_{(i,j)\in\beta_k} \Big(w_H(\xi_i) - w_H(\xi_j) \Big)^2 \le c\Big(1 + \ln \frac{H}{h}\Big) A_k(w_H, w_H), \tag{5.3.11}$$

$$\langle l^{\frac{1}{2}}_{\Gamma_{ij}}\overline{(w_E)}_{ij}, \overline{(w_E)}_{ij}\rangle_{\Gamma_{ij}} \le c\Big(1+\ln\frac{H}{h}\Big)^2 A_k(w_H, w_H),\ \forall\ (i,j)\in\beta_k$$

Thus, the left hand side of (5.3.10) is true. □

Now, we outline the steps for obtaining the solution of

$$w \in V_h^0:\ B_1(w,v) = (g,v),\ \forall\ v\in V_h^0 \tag{5.3.12}$$

Algorithm 5.3.1.

Step 1. Find w_p by solving in parallel the homogeneous Dirichlet subproblems:

$$\begin{cases} w_p \in V_h^{1,0}\oplus V_h^{2,0}\oplus\cdots\oplus V_h^{N,0} \\ A(w_p,v) = (g,v),\ \forall\ v\in V_h^{1,0}\oplus V_h^{2,0}\oplus\cdots\oplus V_h^{N,0} \end{cases}$$

Step 2. Find $\overline{(w_E)}_{ij}$ by solving in parallel one–dimensional equations on each Γ_{ij}:

$$\langle l^{\frac{1}{2}}_{\Gamma_{ij}}\overline{(w_E)}_{ij}, \overline{(v_E)}_{ij}\rangle_{\Gamma_{ij}} = (g,v) - A(w_p,v),$$

for any $v\in V_h^0$ such that $v(x)=0$ if the interpolation point $x\notin\Gamma_{ij}$.

Step 3. Find the values of w_V at $\{\xi_j\}_{j=1}^m$ by solving the coarse mesh problem:

$$A(w_V,v) = (g,v) - A(w_p,v),$$

for any $v\in V_h^0$ such that $v(x)$ is piecewise linear (bilinear) on Ω_h.

Step 4. Find w_H by solving in parallel Dirichlet subproblems (5.3.5) since the values of w_H at the interpolation points on $\bigcup_{k=1}^{N}\partial\Omega_k\backslash\partial\Omega$ follows from Step 2 and Step 3.

Step 5. Set $w = w_p + w_H$.

Besides, we note that Step 2 and Step 3 can be performed simultaneously. Hence, (5.3.12) is easily solved in parallel. Theorem 5.3.2 indicates that $B_1(\cdot,\cdot)$ can decrease the condition number of the system (5.1.2) from $O(h^{-2})$ to $O\Big((1+\ln\frac{H}{h})^2\Big)$. Therefore, $B_1(\cdot,\cdot)$ is an efficient preconditioner.

Suppose $\Gamma_{ij}\subset\partial\Omega_k$. Let $\{\phi_l^k\}$ be the set of the bases of $V_h^{k,0}$. Let $\{\phi_n^{\Gamma_{ij}}\}$ be the set of the bases related to the interpolation points on Γ_{ij}. Denote

$$K = \Big(A_k(\phi_l^k,\phi_n^k)\Big),\qquad K_\Gamma = \Big(A_k(\phi_l^k,\phi_n^{\Gamma_{ij}})\Big),$$
$$K_{\Gamma\Gamma} = \Big(A_k(\phi_l^{\Gamma_{ij}},\phi_n^{\Gamma_{ij}})\Big),\quad S_{ij}^k = K_{\Gamma\Gamma} - K_\Gamma^T K^{-1} K_\Gamma.$$

We call S_{ij}^k the capacitance matrix of Ω_k concerning Γ_{ij}. Lemma 5.2.1 indicates that $\langle l^{\frac{1}{2}}_{\Gamma_{ij}}\overline{(w_E)}_{ij}, \overline{(w_E)}_{ij}\rangle_{\Gamma_{ij}}$ is equivalent to $A_k(\widehat{(w_E)}_{ij}, \widehat{(w_E)}_{ij})$. Thus, we have

Corollary 5.3.3. *If the second term of the right hand side of (5.3.9) is replaced by*

$$\sum_{\Gamma_{ij}} A_k(\widehat{(v_E)}_{ij}, \widehat{(w_E)}_{ij}) = \sum_{\Gamma_{ij}} (v_E)_{ij}^T S_{ij}^k (w_E)_{ij}, \tag{5.3.13}$$

then (5.3.10) is still true.

Remark 5.3.1. If the second term of the right hand side of (5.3.9) is replaced by (5.3.13), only Step 2 of Algorithm 5.3.1 is different. In this case, Step 2 becomes: Find $\widehat{(w_E)}_{ij}$ or $(w_E)_{ij}$ by solving simultaneously the discrete harmonic subproblems with the homogeneous Dirichlet condition on $\partial\Omega_k \backslash \Gamma_{ij}$ and the Neumann condition on Γ_{ij}.

Remark 5.3.2. In (3.5.2), (3.5.3) and (3.5.4), three matrices D, G, and J are introduced, which are all inverted easily by FFT and spectrally equivalent to the capacitance matrix. For Γ_{ij}, they are denoted D_{ij}, G_{ij} and J_{ij}, respectively. Therefore, if the second term of the right hand side of (5.3.9) is replaced by

$$\sum_{\Gamma_{ij}} (v_E)_{ij}^T D_{ij}(w_E)_{ij}, \quad \sum_{\Gamma_{ij}} (v_E)_{ij}^T G_{ij}(w_E)_{ij}, \text{ or } \sum_{\Gamma_{ij}} (v_E)_{ij}^T J_{ij}(w_E)_{ij},$$

the only difference is that Step 2 of Algorithm 5.3.1 can be performed by FFT.

Theorem 5.3.4. *Suppose that* $\{\Omega_k\}_{k=1}^N$ *satisfies the red–black subdivision. Denote*

$$N_R = \Big\{k: \ \Omega_k \text{ is red}, \ 1 \le k \le N\Big\}, \quad N_B = \Big\{k: \ \Omega_k \text{ is black}, \ 1 \le k \le N\Big\}.$$

Then the bilinear form

$$\begin{aligned} B_2(v,w) = \ & A(v_p, w_p) + A(I_H v, I_H w) \\ & + \Big(1 + \ln \tfrac{H}{h}\Big) \textstyle\sum_{k \in N_R} A_k(v_H - I_H v, w_H - I_H w), \end{aligned} \tag{5.3.14}$$

satisfies

$$c\Big(1 + \ln \frac{H}{h}\Big)^{-2} B_2(w,w) \le A(w,w) \le C\Big(1 + \ln \frac{H}{h}\Big) B_2(w,w), \ \forall\, w \in V_h^0 \tag{5.3.15}$$

Proof. Let Ω_k be a black subdomain, $\Gamma_{ij} \subset \partial\Omega_k$. The red subdomain which has the common edge Γ_{ij} with Ω_k is denoted $\Omega_{k_{ij}}$. Since $w_H - I_H w$ is discrete harmonic on Ω_k, it follows from Lemma 5.2.2 1), Lemma 5.2.4 that

$$\begin{aligned} A_k(w_H - I_H w, w_H - I_H w) \ & \le c \sum_{(i,j) \in \beta_k} \langle l_{\Gamma_{ij}}^{\frac{1}{2}} \overline{(w_H - I_H w)_{ij}}, \overline{(w_H - I_H w)_{ij}} \rangle_{\Gamma_{ij}} \\ & \le c\Big(1 + \ln \tfrac{H}{h}\Big)^2 \sum_{(i,j) \in \beta_k} A_{k_{ij}}(w_H - I_H w, w_H - I_H w). \end{aligned}$$

Therefore, the right hand side of (5.3.15) follows from the triangle inequality.

On the other hand, by (5.3.11) and the triangle inequality, we see that the left hand side of (5.3.15) is true. □

Theorem 5.3.5. *If the assumption of Theorem 5.3.4 is satisfied, and if the second term of the right hand side of (5.3.14) is replaced by*

$$(1 + \ln \frac{H}{h}) \sum_{k \in N_R} A_k(v_E, w_E),$$

then (5.3.15) holds still.

The proof of Theorem 5.3.5 is similar to that of Theorem 5.3.4, and omitted here. $B_2(\cdot,\cdot)$ can reduce the condition number of the system (5.1.2) to $O\Big((1+\ln\frac{H}{h})^3\Big)$, and its preconditioning algorithm is different from Algorithm 5.3.1 only in Step 2, where solving in parallel the subproblems on the red subdomains takes the place of solving simultaneously the one–dimensional equations on each Γ_{ij}. Hence, $B_2(\cdot,\cdot)$ is also an efficient preconditioner.

Remark 5.3.3. The term $A(I_H v, I_H w)$ in $B_1(\cdot,\cdot)$ and $B_2(\cdot,\cdot)$ can be replaced by

$$\sum_{\Gamma_{ij}} \alpha_{ij}\Big(v(\xi_i)-v(\xi_j)\Big)\Big(w(\xi_i)-w(\xi_j)\Big) \quad \text{or} \quad A(v_V, w_V),$$

where α_{ij} is some proper positive constant (cf. [12] for details).

Remark 5.3.4. As in [12,25], It is not difficult to write out the matrix representations of $B_1(\cdot,\cdot)$ and $B_2(\cdot,\cdot)$.

Chapter 6

An Iterative Algorithm with Multi–subdomains for the Second Kind Nonconforming Elements ——in the Red–black Subdivision Case

6.1 Introduction

Consider the following elliptic problem:

$$u \in H_0^1(\Omega): \ a(u,v) = (f,v), \ \forall \ v \in H_0^1(\Omega) \tag{6.1.1}$$

where $a(u,v) = \int_\Omega \left[\sum_{i,j=1}^{2} a_{ij}(x) \frac{\partial u}{\partial x_i} \frac{\partial v}{\partial x_j} + a_0(x) uv \right], \ (f,v) = \int_\Omega fv, \ f \in H^{-1}(\Omega),$
$\Omega \subset \Re^2$ is a bounded polygonal open domain , $a_0(x) \geq \alpha > 0, \ a_0(x), \ a_{ij}(x),$ $(i,j = 1,2)$ are piecewise smooth and bounded functions in Ω, (a_{ij}) is a symmetric, uniformly positive definite matrix in Ω. The aim of this chapter is to extend a two–subdomain nonoverlap domain decomposition algorithm[89] (see also Algorithm 3.4.3) to the multi–subdomain cases to solve (6.1.1), which is discretized by the nonconforming finite elements of the second kind. Let $\Omega_h = \{e\}$ be a quasi–uniform mesh of Ω with h as its mesh parameter and V_h the second kind nonconforming finite element space defined in Chapter 1. The nonconforming element discretization of (6.1.1) is

$$u_h \in V_h^0: \ A(u_h, v_h) = (f, v_h), \ \ \forall \ v_h \in V_h^0, \tag{6.1.2}$$

where $$V_h^0 = \{v_h \in V_h: \ v_h(x) = 0, \ \forall \text{ edge midpoint } x \in \partial\Omega\},$$

$$A(u,v) = \sum_{e \subset \Omega} \int_e \left[\sum_{i,j=1}^{2} a_{ij} \frac{\partial u}{\partial x_i} \frac{\partial v}{\partial x_j} + a_0 uv \right].$$

Suppose that $\{\Omega_i\}_{i=1}^N \triangleq \Omega_H$ is a nonoverlap subdivision of Ω, which is, in fact, a quasi–uniform mesh of Ω with H as its mesh parameter. Ω_i is either a quadrilateral or a triangle, referred to as a subdomain. Assume that Ω_h can be obtained by refining Ω_H, so that Ω_H and Ω_h form a two–level triangulation on Ω. Generally, $H \gg h$, therefore, Ω_H and Ω_h are called the coarse mesh and the fine mesh, respectively. Assume that $\{\Omega_i\}_{i=1}^N$ satisfies the red–black subdivision: if all the subdomains are colored by two colors, red and black, then no pair of subdomains which have a common edge have the same color. In the conforming finite element discrete cases, we note that even if the red–black subdivision is satisfied, none of the two–subdomain nonoverlap algorithms in Chapter 3 can be extended to the multi–subdomain cases for the existence of the internal cross points (see [89] and Widlund's paper in [36]). Several algorithms are developed by the introduction of a Lagrangian multiplier or a small scale overlap subdomain in the neighborhood of each internal cross point. But, in the nonconforming discrete cases, it is unnecessary and in fact impossible to calculate the values at the internal cross points. Therefore, we can extend two–subdomain algorithms. The algorithm developed to solve (6.1.2) in this chapter doesn't handle the internal cross points, which plus the simple way to interchange the information between the subdomains leads to its high parallelism; see [48]. In each iteration of our algorithm, the Dirichlet subproblems set on Ω_i are solved simultaneously, which is then followed by the parallel solving of the Neumann subproblems posed on Ω_i. Another remarkable feature of the algorithm is that it is easy to implement. Based on Theorem 2.4.6, it is proved to be geometrically convergent. Virtually, it is equivalent to the simple iterative method applied to the preconditioned capacitance equation. Also, the preconditioner implied in the iteration is easily invertible and the condition number of the preconditioned capacitance matrix is no more than $O((1+\ln\frac{H}{h})(1+H^{-2}+\ln\frac{H}{h}))$.

The remainder is organized as follows: Sect.6.2 describes the domain decomposition algorithm. Sect.6.3 and Sect.6.4 give the convergence analysis and the matrix analysis, respectively. Finally, Sect.6.5 presents numerical experiments to illustrate the efficiency of our algorithm.

6.2 Domain decomposition algorithm

Denote $\Gamma = \bigcup_{i=1}^{N} \partial\Omega_i \backslash \partial\Omega$. Let $\{\xi_j\}_{j=1}^{m}$ be the set of the edge midpoints on Γ (ordered in some way), and $\{\phi_i^\Gamma\}$ the set of the corresponding bases. The discrete trace operator $r_0^h : V_h^0 \to \Re^m$ is defined as follows

$$\forall\, v \in V_h^0,\ r_0^h v \in \Re^m :\ \left(r_0^h v\right)(j) = v(\xi_j),\ j = 1, 2, \cdots, m.$$

Let $\beta_i = \{k :\ \Omega_k \in \Omega_H,\ \mathrm{meas}(\partial\Omega_k \cap \partial\Omega_i) > 0\}$. For $j \in \beta_i$, the open common edge of Ω_i and Ω_j is labeled by Γ_{ij}. For $v \in V_h^0$, $\overline{v}_{ij}$, the piecewise linear continuous function on Γ_{ij}, satisfies

$$\overline{v}_{ij}(\xi_k) = v(\xi_k),\ \forall\, \xi_k \in \Gamma_{ij},\ \overline{v}_{ij}(\nu) = 0,\ \forall \text{ endpoint } \nu \text{ of } \Gamma_{ij}.$$

And, we introduce the following notations:

$$V_h^i = \{v \in V_h^0 :\ v(x) = 0,\ \forall \text{ interpolation point } x \in \Omega\backslash\overline{\Omega}_i\},$$

$$V_h^{i,0} = \{v \in V_h^0 :\ v(x) = 0,\ \forall \text{ interpolation point } x \in \Omega\backslash\Omega_i\},$$

$$A_i(u,v) = \sum_{e\subset\Omega_i} \int_e \left[\sum_{k,j=1}^{2} a_{kj}\frac{\partial u}{\partial x_k}\frac{\partial v}{\partial x_j} + a_0 uv\right],\quad (f,v)_i = \int_{\Omega_i} fv.$$

Here, an interpolation point $x \in \overline{\Omega}$ is related to the definition of V_h.

Denote

$$\Upsilon = \{k : \Omega_k \in \Omega_H, \ \Omega_k \text{ is red}\}, \quad B = \{k : \Omega_k \in \Omega_H, \ \Omega_k \text{ is black}\}.$$

Obviously, we have $\Upsilon \bigcup B = \{1, 2, \cdots, N\}$, and $\beta_k \subset \Upsilon$ for $k \in B$, while $\beta_i \subset B$ for $i \in \Upsilon$. In addition, we introduce the following notations.

$$\begin{aligned}
\Omega_\Upsilon &= \bigcup_{k \in \Upsilon} \Omega_k, & (w, v)_\Upsilon &= \sum_{k \in \Upsilon} (w, v)_k, \\
V_h^\Upsilon &= \bigcup_{k \in \Upsilon} V_h^k, & A_\Upsilon(w, v) &= \sum_{k \in \Upsilon} A_k(w, v), \\
V_h^{\Upsilon,0} &= \bigcup_{k \in \Upsilon} V_h^{k,0}, & D_\Upsilon(v) &= \Big((A_\Upsilon(v, \phi_i^\Gamma) - (f, \phi_i^\Gamma)_\Upsilon)\Big).
\end{aligned}$$

Correspondingly, we have $\Omega_B, V_h^B, V_h^{B,0}, (\cdot,\cdot)_B, A_B(\cdot,\cdot), D_B(\cdot)$. Now, the iterative substructuring method for solving (6.1.2) is stated as follows.

Algorithm 6.2.1. Choose properly the relaxation factors $\theta_1, \theta_2 \in (0, 1)$. Let $g^0 \in \Re^m$ be any initial guess. Suppose that $g^n \in \Re^m$ is obtained. The procedure to get $g^{n+1} \in \Re^m$ consists of the following four steps:

Step 1. For $i = 1, 2, \cdots, N$, solve in parallel

$$\begin{cases} u_i^{2n} \in V_h^i, \\ A_i(u_i^{2n}, v) = (f, v)_i, \quad \forall\, v \in V_h^{i,0}, \\ u_i^{2n}(\xi_j) = \Big(g^n\Big)(j), \quad \forall\, \xi_j \in \partial\Omega_i \backslash \partial\Omega. \end{cases}$$

Step 2. Calculate $u_\Upsilon^{2n} = \sum_{k \in \Upsilon} u_k^{2n}, \quad u_B^{2n} = \sum_{k \in B} u_k^{2n},$

$$d^n = \theta_1 D_\Upsilon(u_\Upsilon^{2n}) - (1 - \theta_1) D_B(u_B^{2n}).$$

Step 3. For $i = 1, 2, \cdots, N$, solve in parallel

$$\begin{cases} i \in \Upsilon, \\ u_i^{2n+1} \in V_h^i, \\ A_i(u_i^{2n+1}, v) = (f, v)_i, & \forall\, v \in V_h^{i,0}, \\ A_i(u_i^{2n+1}, \phi_j^\Gamma) = (f, \phi_j^\Gamma)_i + \Big(d^n\Big)(j), & \forall\, \xi_j \in \partial\Omega_i \backslash \partial\Omega. \end{cases}$$

$$\begin{cases} i \in B, \\ u_i^{2n+1} \in V_h^i, \\ A_i(u_i^{2n+1}, v) = (f, v)_i, & \forall\, v \in V_h^{i,0}, \\ A_i(u_i^{2n+1}, \phi_j^\Gamma) = (f, \phi_j^\Gamma)_i - \Big(d^n\Big)(j), & \forall\, \xi_j \in \partial\Omega_i \backslash \partial\Omega. \end{cases}$$

Step 4. Calculate $u_\Upsilon^{2n+1} = \sum_{k \in \Upsilon} u_k^{2n+1}, \quad u_B^{2n+1} = \sum_{k \in B} u_k^{2n+1},$

$$g^{n+1} = \theta_2 r_0^h u_\Upsilon^{2n+1} + (1 - \theta_2) r_0^h u_B^{2n+1}.$$

6.3 Convergence analysis

Lemma 6.3.1. *Let* $v \in V_h^0$ *be the discrete harmonic function on* Ω_i, *i.e.,* v *satisfies*

$$A_i(v,\theta) = 0, \quad \forall\, \theta \in V_h^{i,0}$$

Then there exists a positive constant c, independent of h, H *and* Ω_i, *such that*

$$A_i(v,v) \le c \sum_{j\in\beta_i} \|\overline{v}_{ij}\|^2_{H_{00}^{\frac{1}{2}}(\Gamma_{ij})}.$$

Proof. Let $w_j \in V_h^0$ such that

$$\begin{cases} A_i(w_j,\theta) = 0, & \forall\, \theta \in V_h^{i,0} \\ w_j(x) = v(x), & \forall \text{ interpolation point } x \in \Gamma_{ij} \\ w_j(x) = 0, & \forall \text{ interpolation point } x \in \partial\Omega_i \backslash \Gamma_{ij} \end{cases}$$

Then

$$\sum_{j\in\beta_i} w_j = v, \text{ on } \Omega_i$$

By triangle inequality and Theorem 2.4.3, we know Lemma 6.3.1 is true. □

Let $E_\Upsilon^h : \Re^m \to V_h^\Upsilon$ and $E_B^h : \Re^m \to V_h^B$ be nonconforming discrete harmonic extension operators, which satisfies, respectively, for any $\lambda \in \Re^m$

$$\begin{cases} E_\Upsilon^h\lambda \in V_h^\Upsilon, \\ A_\Upsilon(E_\Upsilon^h\lambda, v) = 0, \ \forall\, v \in V_h^{\Upsilon,0} \\ r_0^h E_\Upsilon^h\lambda = \lambda, \end{cases}, \qquad \begin{cases} E_B^h\lambda \in V_h^B, \\ A_B(E_B^h\lambda, v) = 0, \ \forall\, v \in V_h^{B,0}, \\ r_0^h E_B^h\lambda = \lambda. \end{cases}$$

Furthermore, $\forall\, \lambda \in \Re^m$, the extension operators $T_\Upsilon^h : \Re^m \to V_h^\Upsilon$ and $T_B^h : \Re^m \to V_h^B$ are defined as follows

$$\begin{cases} T_\Upsilon^h\lambda \in V_h^\Upsilon, \\ A_\Upsilon(T_\Upsilon^h\lambda, v) = 0, & \forall\, v \in V_h^{\Upsilon,0}, \\ A_\Upsilon(T_\Upsilon^h\lambda, \phi_i^\Gamma) = -A_B(E_B^h\lambda, \phi_i^\Gamma), & i = 1,2,\cdots,m. \end{cases}$$

$$\begin{cases} T_B^h\lambda \in V_h^B, \\ A_B(T_B^h\lambda, v) = 0, & \forall\, v \in V_h^{B,0}, \\ A_B(T_B^h\lambda, \phi_i^\Gamma) = -A_\Upsilon(E_\Upsilon^h\lambda, \phi_i^\Gamma), & i = 1,2,\cdots,m. \end{cases}$$

For $\lambda \in \Re^m$, it is easy to see that

$$A_\Upsilon(T_\Upsilon^h\lambda, v) = -A_B(E_B^h\lambda, v), \quad \forall\, v \in V_h, \tag{6.3.1}$$

$$A_B(T_B^h\lambda, v) = -A_\Upsilon(E_\Upsilon^h\lambda, v), \quad \forall\, v \in V_h, \tag{6.3.2}$$

$$A_\Upsilon(T_\Upsilon^h\lambda, v) = -A_B(E_B^h\lambda, E_B^h r_0^h v), \quad \forall\, v \in V_h^\Upsilon, \tag{6.3.3}$$

$$A_B(T_B^h\lambda, v) = -A_\Upsilon(E_\Upsilon^h\lambda, E_\Upsilon^h r_0^h v), \quad \forall\, v \in V_h^B. \tag{6.3.4}$$

Theorem 6.3.2. (The extension theorem in multi-subdomain cases)

$$\sigma = \sup_{\lambda\in\Re^m\setminus\{0\}} \frac{A_\Upsilon(E_\Upsilon^h\lambda, E_\Upsilon^h\lambda)}{A_B(E_B^h\lambda, E_B^h\lambda)} \le c(1+\ln\frac{H}{h})(1+\ln\frac{H}{h}+H^{-2}), \tag{6.3.5}$$

$$\tau = \sup_{\lambda\in\Re^m\setminus\{0\}} \frac{A_B(E_B^h\lambda, E_B^h\lambda)}{A_\Upsilon(E_\Upsilon^h\lambda, E_\Upsilon^h\lambda)} \le c(1+\ln\frac{H}{h})(1+\ln\frac{H}{h}+H^{-2}). \tag{6.3.6}$$

Proof. $\forall\, \lambda \in \Re^m$, let $v = \begin{cases} E_B^h\lambda, & \text{on } \overline{\Omega}_B \\ E_\Upsilon^h\lambda, & \text{on } \Omega_\Upsilon \end{cases}$. It follows from Lemma 6.3.1 and Lemma 2.4.6 that

$$\begin{aligned} A_i(v,v) &\le c\sum_{j\in\beta_i} \|\overline{v}_{ij}\|^2_{H_{00}^{\frac{1}{2}}(\Gamma_{ij})} \\ &\le c\sum_{j\in\beta_i}(1+\ln\frac{H}{h})\Big\{(1+\ln\frac{H}{h})|v|^2_{1,\Omega_j,h} + (1+H^{-2})\|v\|^2_{L^2(\Omega_j)}\Big\} \\ &\le c(1+\ln\frac{H}{h})(1+H^{-2}+\ln\frac{H}{h})\sum_{j\in\beta_i} A_j(v,v). \end{aligned}$$

Summing up with $i \in \Upsilon$ yields

$$A_\Upsilon(v,v) \le c(1+\ln\frac{H}{h})(1+H^{-2}+\ln\frac{H}{h})A_B(v,v),$$

where we have applied the fact that $\beta_i \subset B$ for $i \in \Upsilon$. Therefore, (6.3.5) holds.

Similarly, (6.3.6) can be established . This completes the proof. □

Corollary 6.3.3. *Let σ and τ be those in Theorem 6.3.2. We have*

$$\frac{1}{\tau}A_\Upsilon(E_\Upsilon^h\lambda, E_\Upsilon^h\lambda) \le A_B(T_B^h\lambda, T_B^h\lambda) \le \sigma A_\Upsilon(E_\Upsilon^h\lambda, E_\Upsilon^h\lambda), \quad \forall\, \lambda \in \Re^m \tag{6.3.7}$$

$$\frac{1}{\sigma}A_B(E_B^h\lambda, E_B^h\lambda) \le A_\Upsilon(T_\Upsilon^h\lambda, T_\Upsilon^h\lambda) \le \tau A_B(E_B^h\lambda, E_B^h\lambda), \quad \forall\, \lambda \in \Re^m \tag{6.3.8}$$

Proof. Taking $v = T_B^h\lambda$ in (6.3.4), using the Schwarz's inequality and (6.3.5), we obtain

$$\begin{aligned} A_B(T_B^h\lambda, T_B^h\lambda) &= -A_\Upsilon(E_\Upsilon^h\lambda, E_\Upsilon^h r_0^h T_B^h\lambda) \\ &\le \left(A_\Upsilon(E_\Upsilon^h\lambda, E_\Upsilon^h\lambda)\right)^{\frac{1}{2}}\left(A_\Upsilon(E_\Upsilon^h r_0^h T_B^h\lambda, E_\Upsilon^h r_0^h T_B^h\lambda)\right)^{\frac{1}{2}} \\ &\le \left(A_\Upsilon(E_\Upsilon^h\lambda, E_\Upsilon^h\lambda)\right)^{\frac{1}{2}}\left(\sigma A_B(T_B^h\lambda, T_B^h\lambda)\right)^{\frac{1}{2}} \end{aligned}$$

which implies that the right hand side of (6.3.7) holds.

On the other hand, the substitution of v in (6.3.4) with $E_B^h\lambda$ and (6.3.6) yield

$$A_\Upsilon(E_\Upsilon^h\lambda, E_\Upsilon^h\lambda) = -A_B(T_B^h\lambda, E_B^h\lambda) \leq \left(A_B(T_B^h\lambda, T_B^h\lambda)\right)^{\frac{1}{2}}\left(A_B(E_B^h\lambda, E_B^h\lambda)\right)^{\frac{1}{2}}$$
$$\leq \left(A_B(T_B^h\lambda, T_B^h\lambda)\right)^{\frac{1}{2}}\left(\tau A_\Upsilon(E_\Upsilon^h\lambda, E_\Upsilon^h\lambda)\right)^{\frac{1}{2}}$$

which indicates that the left hand side of (6.3.7) is true. And we get (6.3.7).

(6.3.8) follows in the same manner. Hence the proof is complete. □

Theorem 6.3.4. *Let u_h be the solution of (6.1.2), and u^n be the approximate solution obtained by Algorithm 6.2.1, where $u^n = u_i^n$ in Ω_i. Denote $\varepsilon^n = u_h - u^n$. Then, we have*

1) There exist θ_1^* and $\theta_2^* \in (0,1)$ such that if $\theta_1 \in (\theta_1^*, 1)$ and $\theta_2 \in (\theta_2^*, 1)$, then

$$A(\varepsilon^{n+2}, \varepsilon^{n+2}) \leq \kappa_1(\theta_1)\kappa_2(\theta_2)A(\varepsilon^n, \varepsilon^n), \tag{6.3.9}$$

with $0 \leq \kappa_1(\theta_1) < 1$ and $0 \leq \kappa_2(\theta_2) < 1$.

2) There exist the optimal relaxation factors θ_1^{opt} and θ_2^{opt}, such that

$$\kappa_1(\theta_1^{opt})\kappa_2(\theta_2^{opt}) = \min_{\theta_1 \in (\theta_1^*, 1)} \min_{\theta_2 \in (\theta_2^*, 1)} \kappa_1(\theta_1)\kappa_2(\theta_2). \tag{6.3.10}$$

Proof. Let $u_h^i \in V_h^i$ s.t.

$$u_h^i(x) = u_h(x), \ \forall \text{ interpolation point } x \in \overline{\Omega}_i.$$

Denote

$$u_h^\Upsilon = \sum_{k\in\Upsilon} u_h^k \in V_h^\Upsilon, \qquad u_h^B = \sum_{k\in B} u_h^k \in V_h^B,$$
$$\varepsilon_i^n = u_i^n - u_h^i, \qquad \mu^n = g^n - r_0^h u_h,$$
$$\varepsilon_\Upsilon^n = u_\Upsilon^n - u_h^\Upsilon, \qquad \varepsilon_B^n = u_B^n - u_h^B.$$

It is easy to see that

$$\begin{cases} \varepsilon_\Upsilon^{2n} \in V_h^\Upsilon, \\ A_\Upsilon(\varepsilon_\Upsilon^{2n}, v) = 0, \quad \forall\, v \in V_h^{\Upsilon,0}, \\ r_0^h \varepsilon_\Upsilon^{2n} = \mu^n. \end{cases} \tag{6.3.11}$$

$$\begin{cases} \varepsilon_B^{2n} \in V_h^B, \\ A_B(\varepsilon_B^{2n}, v) = 0, \quad \forall\, v \in V_h^{B,0}, \\ r_0^h \varepsilon_B^{2n} = \mu^n. \end{cases} \tag{6.3.12}$$

$$\left(d^n\right)(i) = \theta_1 A_\Upsilon(\varepsilon_\Upsilon^{2n}, \phi_i^\Gamma) - (1-\theta_1)A_B(\varepsilon_B^{2n}, \phi_i^\Gamma), \ i = 1, 2, \cdots, m. \tag{6.3.13}$$

$$\begin{cases} \varepsilon_\Upsilon^{2n+1} \in V_h^\Upsilon, \\ A_\Upsilon(\varepsilon_\Upsilon^{2n+1}, v) = 0, \quad \forall\, v \in V_h^{\Upsilon,0}, \\ A_\Upsilon(\varepsilon_\Upsilon^{2n+1}, \phi_i^\Gamma) = \left(d^n\right)(i), \quad i = 1, 2, \cdots, m. \end{cases} \tag{6.3.14}$$

$$\begin{cases} \varepsilon_B^{2n+1} \in V_h^B, \\ A_B(\varepsilon_B^{2n+1}, v) = 0, & \forall\, v \in V_h^{B,0}, \\ A_B(\varepsilon_B^{2n+1}, \phi_i^\Gamma) = -\Big(d^n\Big)(i), & i = 1, 2, \cdots, m. \end{cases} \tag{6.3.15}$$

$$\mu^{n+1} = \theta_2 r_0^h \varepsilon_\Upsilon^{2n+1} + (1-\theta_2) r_0^h \varepsilon_B^{2n+1}. \tag{6.3.16}$$

By (6.3.14), we have

$$A_\Upsilon(\varepsilon_\Upsilon^{2n+1}, \phi_i^\Gamma) = \theta_1 A_\Upsilon(\varepsilon_\Upsilon^{2n}, \phi_i^\Gamma) - (1-\theta_1) A_B(\varepsilon_B^{2n}, \phi_i^\Gamma).$$

Hence

$$\varepsilon_\Upsilon^{2n+1} = \theta_1 \varepsilon_\Upsilon^{2n} + (1-\theta_1) T_\Upsilon^h r_0^h \varepsilon_B^{2n} = \theta_1 \varepsilon_\Upsilon^{2n} + (1-\theta_1) T_\Upsilon^h r_0^h \varepsilon_\Upsilon^{2n} \tag{6.3.17}$$

$$\begin{aligned} A_\Upsilon(\varepsilon_\Upsilon^{2n+1}, \varepsilon_\Upsilon^{2n+1}) = \;& \theta_1^2 A_\Upsilon(\varepsilon_\Upsilon^{2n}, \varepsilon_\Upsilon^{2n}) + (1-\theta_1)^2 A_\Upsilon(T_\Upsilon^h r_0^h \varepsilon_\Upsilon^{2n}, T_\Upsilon^h r_0^h \varepsilon_\Upsilon^{2n}) \\ & + 2\theta_1(1-\theta_1) A_\Upsilon(\varepsilon_\Upsilon^{2n}, T_\Upsilon^h r_0^h \varepsilon_\Upsilon^{2n}) \end{aligned} \tag{6.3.18}$$

It follows from (6.3.3) and (6.3.5) that

$$\begin{aligned} A_\Upsilon(T_\Upsilon^h r_0^h \varepsilon_\Upsilon^{2n}, \varepsilon_\Upsilon^{2n}) &= -A_B(E_B^h r_0^h \varepsilon_\Upsilon^{2n}, E_B^h r_0^h \varepsilon_\Upsilon^{2n}) \\ &\le -\frac{1}{\sigma} A_\Upsilon(E_\Upsilon^h r_0^h \varepsilon_\Upsilon^{2n}, E_\Upsilon^h r_0^h \varepsilon_\Upsilon^{2n}) = -\frac{1}{\sigma} A_\Upsilon(\varepsilon_\Upsilon^{2n}, \varepsilon_\Upsilon^{2n}). \end{aligned} \tag{6.3.19}$$

(6.3.8) and (6.3.6) yield

$$A_\Upsilon(T_\Upsilon^h r_0^h \varepsilon_\Upsilon^{2n}, T_\Upsilon^h r_0^h \varepsilon_\Upsilon^{2n}) \le \tau A_B(E_B^h r_0^h \varepsilon_\Upsilon^{2n}, E_B^h r_0^h \varepsilon_\Upsilon^{2n}) \le \tau^2 A_\Upsilon(\varepsilon_\Upsilon^{2n}, \varepsilon_\Upsilon^{2n}). \tag{6.3.20}$$

With the substitution of (6.3.19) and (6.3.20) into (6.3.18), we get

$$A_\Upsilon(\varepsilon_\Upsilon^{2n+1}, \varepsilon_\Upsilon^{2n+1}) \le \kappa_1(\theta_1) A_\Upsilon(\varepsilon_\Upsilon^{2n}, \varepsilon_\Upsilon^{2n}), \tag{6.3.21}$$

where

$$\kappa_1(\theta_1) = \frac{1}{\sigma}\Big((1-\theta_1)^2(\sigma\tau^2 + \sigma + 2) - 2(1-\theta_1)(\sigma+1) + \sigma\Big). \tag{6.3.22}$$

On the other hand, it follows from (6.3.17) that

$$E_B^h r_0^h \varepsilon_\Upsilon^{2n+1} = \theta_1 E_B^h r_0^h \varepsilon_\Upsilon^{2n} + (1-\theta_1) E_B^h r_0^h T_\Upsilon^h r_0^h \varepsilon_\Upsilon^{2n} = \theta_1 \varepsilon_B^{2n} + (1-\theta_1) E_B^h r_0^h T_\Upsilon^h r_0^h \varepsilon_\Upsilon^{2n},$$

$$\begin{aligned} A_B(E_B^h r_0^h \varepsilon_\Upsilon^{2n+1}, E_B^h r_0^h \varepsilon_\Upsilon^{2n+1}) = \;& \theta_1^2 A_B(\varepsilon_B^{2n}, \varepsilon_B^{2n}) + (1-\theta_1)^2 \\ & A_B(E_B^h r_0^h T_\Upsilon^h r_0^h \varepsilon_\Upsilon^{2n}, E_B^h r_0^h T_\Upsilon^h r_0^h \varepsilon_\Upsilon^{2n}) \\ & + 2\theta_1(1-\theta_1) A_B(\varepsilon_B^{2n}, E_B^h r_0^h T_\Upsilon^h r_0^h \varepsilon_\Upsilon^{2n}). \end{aligned} \tag{6.3.23}$$

Applying (6.3.6) and (6.3.8) gives

$$\begin{aligned} A_B(E_B^h r_0^h T_\Upsilon^h r_0^h \varepsilon_\Upsilon^{2n}, E_B^h r_0^h T_\Upsilon^h r_0^h \varepsilon_\Upsilon^{2n}) &\le \tau A_\Upsilon(E_\Upsilon^h r_0^h T_\Upsilon^h r_0^h \varepsilon_\Upsilon^{2n}, E_\Upsilon^h r_0^h T_\Upsilon^h r_0^h \varepsilon_\Upsilon^{2n}) \\ &= \tau A_\Upsilon(T_\Upsilon^h r_0^h \varepsilon_\Upsilon^{2n}, T_\Upsilon^h r_0^h \varepsilon_\Upsilon^{2n}) \\ &\le \tau^2 A_B(E_B^h r_0^h \varepsilon_\Upsilon^{2n}, E_B^h r_0^h \varepsilon_\Upsilon^{2n}) \\ &= \tau^2 A_B(\varepsilon_B^{2n}, \varepsilon_B^{2n}). \end{aligned} \tag{6.3.24}$$

By (6.3.3) and (6.3.8), we obtain

$$\begin{aligned} A_B(\varepsilon_B^{2n}, E_B^h r_0^h T_\Upsilon^h r_0^h \varepsilon_\Upsilon^{2n}) &= A_B(E_B^h r_0^h \varepsilon_B^{2n}, E_B^h r_0^h T_\Upsilon^h r_0^h \varepsilon_B^{2n}) \\ &= A_\Upsilon(T_\Upsilon^h r_0^h \varepsilon_B^{2n}, T_\Upsilon^h r_0^h \varepsilon_B^{2n}) \\ &\le -\frac{1}{\sigma} A_B(E_B^h r_0^h \varepsilon_B^{2n}, E_B^h r_0^h \varepsilon_B^{2n}) \\ &= -\frac{1}{\sigma} A_B(\varepsilon_B^{2n}, \varepsilon_B^{2n}). \end{aligned} \tag{6.3.25}$$

With the substitution of (6.3.24) and (6.3.25) into (6.3.23) , we get

$$A_B(E_B^h r_0^h \varepsilon_\Upsilon^{2n+1}, E_B^h r_0^h \varepsilon_\Upsilon^{2n+1}) \le \kappa_1(\theta_1) A_B(\varepsilon_B^{2n}, \varepsilon_B^{2n}). \tag{6.3.26}$$

From (6.3.11), it is easy to see that

$$\varepsilon_\Upsilon^{2n+2} = \theta_2 E_\Upsilon^h r_0^h \varepsilon_\Upsilon^{2n+1} + (1-\theta_2) E_\Upsilon^h r_0^h \varepsilon_B^{2n+1} = \theta_2 \varepsilon_\Upsilon^{2n+1} + (1-\theta_2) E_\Upsilon^h r_0^h \varepsilon_B^{2n+1}.$$

$$\begin{aligned} A_\Upsilon(\varepsilon_\Upsilon^{2n+2}, \varepsilon_\Upsilon^{2n+2}) = \; & \theta_2^2 A_\Upsilon(\varepsilon_\Upsilon^{2n+1}, \varepsilon_\Upsilon^{2n+1}) \\ & + 2\theta_2(1-\theta_2) A_\Upsilon(\varepsilon_\Upsilon^{2n+1}, E_\Upsilon^h r_0^h \varepsilon_B^{2n+1}) \\ & + (1-\theta_2)^2 A_\Upsilon(E_\Upsilon^h r_0^h \varepsilon_B^{2n+1}, E_\Upsilon^h r_0^h \varepsilon_B^{2n+1}) \end{aligned} \tag{6.3.27}$$

(6.3.5) and (6.3.7) gives

$$\begin{aligned} A_\Upsilon(E_\Upsilon^h r_0^h \varepsilon_B^{2n+1}, E_\Upsilon^h r_0^h \varepsilon_B^{2n+1}) &\le \sigma A_B(E_B^h r_0^h \varepsilon_B^{2n+1}, E_B^h r_0^h \varepsilon_B^{2n+1}) \\ &= \sigma A_B(\varepsilon_B^{2n+1}, \varepsilon_B^{2n+1}) \\ &= \sigma A_B(T_B^h r_0^h \varepsilon_\Upsilon^{2n+1}, T_B^h r_0^h \varepsilon_\Upsilon^{2n+1}) \\ &\le \sigma^2 A_\Upsilon(E_\Upsilon^h r_0^h \varepsilon_\Upsilon^{2n+1}, E_\Upsilon^h r_0^h \varepsilon_\Upsilon^{2n+1}) \\ &= \sigma^2 A_\Upsilon(\varepsilon_\Upsilon^{2n+1}, \varepsilon_\Upsilon^{2n+1}). \end{aligned} \tag{6.3.28}$$

It follows from (6.3.4) and (6.3.7) that

$$\begin{aligned} A_\Upsilon(\varepsilon_\Upsilon^{2n+1}, E_\Upsilon^h r_0^h \varepsilon_B^{2n+1}) &= A_\Upsilon(E_\Upsilon^h r_0^h \varepsilon_\Upsilon^{2n+1}, E_\Upsilon^h r_0^h \varepsilon_B^{2n+1}) \\ &= -A_B(T_B^h r_0^h \varepsilon_\Upsilon^{2n+1}, \varepsilon_B^{2n+1}) \\ &= -A_B(T_B^h r_0^h \varepsilon_\Upsilon^{2n+1}, T_B^h r_0^h \varepsilon_\Upsilon^{2n+1}) \\ &\le -\frac{1}{\tau} A_\Upsilon(E_\Upsilon^h r_0^h \varepsilon_\Upsilon^{2n+1}, E_\Upsilon^h r_0^h \varepsilon_\Upsilon^{2n+1}) \\ &= -\frac{1}{\tau} A_\Upsilon(\varepsilon_\Upsilon^{2n+1}, \varepsilon_\Upsilon^{2n+1}). \end{aligned} \tag{6.3.29}$$

With the substitution of (6.3.28) and (6.3.29) into (6.3.27) , we obtain

$$A_\Upsilon(\varepsilon_\Upsilon^{2n+2}, \varepsilon_\Upsilon^{2n+2}) \le \kappa_2(\theta_2) A_\Upsilon(\varepsilon_\Upsilon^{2n+1}, \varepsilon_\Upsilon^{2n+1}), \tag{6.3.30}$$

where

$$\kappa_2(\theta_2) = \frac{1}{\tau}\Big((1-\theta_2)^2(\sigma^2\tau+\tau+2) - 2(1-\theta_2)(\tau+1)+\tau\Big). \tag{6.3.31}$$

On the other hand, (6.3.12) gives

$$\varepsilon_B^{2n+2} = \theta_2 E_B^h r_0^h \varepsilon_\Upsilon^{2n+1} + (1-\theta_2)\varepsilon_B^{2n+1}$$

$$\begin{aligned} A_B(\varepsilon_B^{2n+2},\varepsilon_B^{2n+2}) = \ & \theta_2^2 A_B(E_B^h r_0^h \varepsilon_\Upsilon^{2n+1}, E_B^h r_0^h \varepsilon_\Upsilon^{2n+1}) \\ & +(1-\theta_2)^2 A_B(\varepsilon_B^{2n+1},\varepsilon_B^{2n+1}) \\ & +2\theta_2(1-\theta_2)A_B(\varepsilon_B^{2n+1}, E_B^h r_0^h \varepsilon_\Upsilon^{2n+1}). \end{aligned} \tag{6.3.32}$$

By (6.3.7) and (6.3.5) , we have

$$\begin{aligned} A_B(\varepsilon_B^{2n+1},\varepsilon_B^{2n+1}) & = A_B(T_B^h r_0^h \varepsilon_\Upsilon^{2n+1}, T_B^h r_0^h \varepsilon_\Upsilon^{2n+1}) \\ & \le \sigma A_\Upsilon(E_\Upsilon^h r_0^h \varepsilon_\Upsilon^{2n+1}, E_\Upsilon^h r_0^h \varepsilon_\Upsilon^{2n+1}) \\ & \le \sigma^2 A_B(E_B^h r_0^h \varepsilon_\Upsilon^{2n+1}, E_B^h r_0^h \varepsilon_\Upsilon^{2n+1}). \end{aligned} \tag{6.3.33}$$

(6.3.4) and (6.3.6) imply

$$\begin{aligned} A_B(E_B^h r_0^h \varepsilon_\Upsilon^{2n+1},\varepsilon_B^{2n+1}) & = A_B(T_B^h r_0^h \varepsilon_\Upsilon^{2n+1}, E_B^h r_0^h \varepsilon_\Upsilon^{2n+1}) \\ & = -A_\Upsilon(E_\Upsilon^h r_0^h \varepsilon_\Upsilon^{2n+1}, E_\Upsilon^h r_0^h E_B^h r_0^h \varepsilon_\Upsilon^{2n+1}) \\ & = -A_\Upsilon(E_\Upsilon^h r_0^h \varepsilon_\Upsilon^{2n+1}, E_\Upsilon^h r_0^h \varepsilon_\Upsilon^{2n+1}) \\ & \le -\frac{1}{\tau} A_B(E_B^h r_0^h \varepsilon_\Upsilon^{2n+1}, E_B^h r_0^h \varepsilon_\Upsilon^{2n+1}). \end{aligned} \tag{6.3.34}$$

The substitution of (6.3.33) and (6.3.34) into (6.3.32) reduces to

$$A_B(\varepsilon_B^{2n+2},\varepsilon_B^{2n+2}) \le \kappa_2(\theta_2) A_B(E_B^h r_0^h \varepsilon_\Upsilon^{2n+1}, E_B^h r_0^h \varepsilon_\Upsilon^{2n+1}). \tag{6.3.35}$$

It follows from (6.3.21), (6.3.26), (6.3.30) and (6.3.35) that

$$A_\Upsilon(\varepsilon_\Upsilon^{2n+2},\varepsilon_\Upsilon^{2n+2}) \le \kappa_1(\theta_1)\kappa_2(\theta_2) A_\Upsilon(\varepsilon_\Upsilon^{2n},\varepsilon_\Upsilon^{2n}), \tag{6.3.36}$$

$$A_B(\varepsilon_B^{2n+2},\varepsilon_B^{2n+2}) \le \kappa_1(\theta_1)\kappa_2(\theta_2) A_B(\varepsilon_B^{2n},\varepsilon_B^{2n}). \tag{6.3.37}$$

Similarly, we obtain that

$$A_\Upsilon(\varepsilon_\Upsilon^{2n+3},\varepsilon_\Upsilon^{2n+3}) \le \kappa_1(\theta_1)\kappa_2(\theta_2) A_\Upsilon(\varepsilon_\Upsilon^{2n+1},\varepsilon_\Upsilon^{2n+1}), \tag{6.3.38}$$

$$A_B(\varepsilon_B^{2n+3},\varepsilon_B^{2n+3}) \le \kappa_1(\theta_1)\kappa_2(\theta_2) A_B(\varepsilon_B^{2n+1},\varepsilon_B^{2n+1}). \tag{6.3.39}$$

By (6.3.22) and (6.3.31), an elementary calculation gives

$$\kappa_1(\theta_1) < 1, \quad \forall\, \theta_1 \in (\theta_1^*, 1) = \Big(\frac{\tau^2\sigma - \sigma}{\tau^2\sigma+\sigma+2}, 1\Big)$$

$$\kappa_2(\theta_2) < 1, \quad \forall\, \theta_2 \in (\theta_2^*, 1) = \Big(\frac{\sigma^2\tau - \tau}{\sigma^2\tau + \tau + 2}, 1\Big)$$

Furthermore, when

$$\theta_1 = \theta_1^{opt} = \frac{\tau^2\sigma + 1}{\tau^2\sigma + \sigma + 2}, \quad \theta_2 = \theta_2^{opt} = \frac{\sigma^2\tau + 1}{\sigma^2\tau + \tau + 2}, \tag{6.3.40}$$

$\kappa_1(\theta_1), \kappa_2(\theta_2)$ attain their minimum values, respectively,

$$(\kappa_1)_{min} = \frac{\tau^2\sigma^2 - 1}{\sigma(\tau^2\sigma + \sigma + 2)}, \quad (\kappa_2)_{min} = \frac{\tau^2\sigma^2 - 1}{\tau(\sigma^2\tau + \tau + 2)}. \tag{6.3.41}$$

By now, the proof of Theorem 6.3.4 is complete. □

6.4 Matrix analysis

Suppose that the interpolation points are ordered in the way: first $\{\xi_j\}_{j=1}^m$, then those in Ω_Υ, and finally those in Ω_B. Let $\{\phi_i^k\}$ be the set of bases of $V_h^{k,0}$. We introduce the following notations

$$\begin{aligned}
&\{\phi_j^I\} = \bigcup_{k=1}^{N}\{\phi_i^k\}, && \{\phi_j^\Upsilon\} = \bigcup_{k\in\Upsilon}\{\phi_i^k\}, && \{\phi_j^B\} = \bigcup_{k\in B}\{\phi_i^k\},\\
&A_{II} = \big(A(\phi_i^I, \phi_j^I)\big), && A_{\Upsilon\Upsilon} = \big(A_\Upsilon(\phi_i^\Upsilon, \phi_j^\Upsilon)\big), && A_{BB} = \big(A_B(\phi_i^B, \phi_j^B)\big),\\
&A_{I\Gamma} = \big(A(\phi_i^I, \phi_j^\Gamma)\big), && A_{\Upsilon\Gamma} = \big(A_\Upsilon(\phi_i^\Upsilon, \phi_j^\Gamma)\big), && A_{B\Gamma} = \big(A_B(\phi_i^B, \phi_j^\Gamma)\big),\\
&A_{\Gamma\Gamma} = \big(A(\phi_i^\Gamma, \phi_j^\Gamma)\big), && A_{\Gamma\Gamma}^\Upsilon = \big(A_\Upsilon(\phi_i^\Gamma, \phi_j^\Gamma)\big), && A_{\Gamma\Gamma}^B = \big(A_B(\phi_i^\Gamma, \phi_j^\Gamma)\big),
\end{aligned}$$

$$\begin{aligned}
S &= A_{\Gamma\Gamma} - A_{I\Gamma}^T A_{II}^{-1} A_{I\Gamma},\\
S^\Upsilon &= A_{\Gamma\Gamma}^\Upsilon - A_{\Upsilon\Gamma}^T A_{\Upsilon\Upsilon}^{-1} A_{\Upsilon\Gamma},\\
S^B &= A_{\Gamma\Gamma}^B - A_{B\Gamma}^T A_{BB}^{-1} A_{B\Gamma},
\end{aligned}$$

where, S, S^Υ and S^B are called the capacitance matrices of Ω, Ω_Υ and Ω_B with respect to Γ, respectively. Similarly, we can write out the capacitance matrix of the red (black) subdomain Ω_k with respect to $\partial\Omega_k\backslash\partial\Omega$, which is denoted by S_k^Υ (S_k^B). Obviously, when $\{\xi_i\}_{i=1}^m$ are ordered properly, S^Υ can be expressed as

$$S^\Upsilon = block - diag(S_k^\Upsilon)_{k\in\Upsilon}. \tag{6.4.1}$$

Supposed that $u_h = \sum_{i=1}^m u_i^\Gamma \phi_i^\Gamma + \sum_j u_j^\Upsilon \phi_j^\Upsilon + \sum_l u_l^B \phi_l^B$ is the solution of (6.1.2). Denote $U_\Gamma = (u_i^\Gamma)$, $U_\Upsilon = (u_j^\Upsilon)$, $U_B = (u_l^B)$. We can rewrite (6.1.2) as

$$\begin{bmatrix} A_{\Gamma\Gamma} & A_{\Upsilon\Gamma}^T & A_{B\Gamma}^T \\ A_{\Upsilon\Gamma} & A_{\Upsilon\Upsilon} & 0 \\ A_{B\Gamma} & 0 & A_{BB} \end{bmatrix} \begin{bmatrix} U_\Gamma \\ U_\Upsilon \\ U_B \end{bmatrix} = \begin{bmatrix} F_\Gamma \\ F_\Upsilon \\ F_B \end{bmatrix}. \tag{6.4.2}$$

By block Gaussian elimination, the system (6.4.2) then reduces to

$$SU_\Gamma = \tilde{F}_\Gamma, \tag{6.4.3}$$

where $\tilde{F}_\Gamma = F_\Gamma - A_{\Upsilon\Gamma}^T A_{\Upsilon\Upsilon}^{-1} F_\Upsilon - A_{B\Gamma}^T A_{BB}^{-1} F_B$. Detailed analysis indicates that Algorithm 6.2.1 is equivalent to the following iterative method

$$g^{n+1} = g^n + \Big\{\theta_2(1-\theta_1)[S^\Upsilon]^{-1} + \theta_1(1-\theta_2)[S^B]^{-1}\Big\}(\tilde{F}_\Gamma - Sg^n), \tag{6.4.4}$$

for the capacitance equation (6.4.3) which has been preconditioned by

$$Q = \{\theta_2(1-\theta_1)[S^\Upsilon]^{-1} + \theta_1(1-\theta_2)[S^B]^{-1}\}^{-1}. \tag{6.4.5}$$

It follows from (6.4.1) that

$$[S^\Upsilon]^{-1} = block - diag\Big([S_k^\Upsilon]^{-1}\Big)_{k\in\Upsilon},$$

which implies that $[S^\Upsilon]^{-1}\lambda$ can be obtained through solving in parallel all the subproblems on the red subdomains. On the other hand, we have

Theorem 6.4.1. *The capacitance matrices S, S^Υ and S^B are symmetric, positive definite, and satisfies*

$$\lambda^T S^\Upsilon \lambda \le \lambda^T S \lambda \le c(1+\ln\frac{H}{h})(1+\ln\frac{H}{h}+H^{-2})\lambda^T S^\Upsilon\lambda, \ \forall\, \lambda \in \Re^m \tag{6.4.6}$$

$$\lambda^T S^B \lambda \le \lambda^T S \lambda \le c(1+\ln\frac{H}{h})(1+\ln\frac{H}{h}+H^{-2})\lambda^T S^B\lambda, \ \forall\, \lambda \in \Re^m \tag{6.4.7}$$

Proof. It is easy to see that

$$\forall\, \lambda,\mu \in \Re^m, \ \ \lambda^T S^\Upsilon \mu = A_\Upsilon(E_\Upsilon^h\lambda, E_\Upsilon^h\mu), \ \ \lambda^T S^B\mu = A_B(E_B^h\lambda, E_B^h\mu).$$

By $S = S^\Upsilon + S^B$ and Theorem 6.3.2, we get (6.4.6) and (6.4.7). □

By now, we note that S^Υ, S^B or Q are, respectively, the efficient preconditioners for S, so is the following matrix

$$\begin{bmatrix} Q + A_{\Upsilon\Gamma}^T A_{\Upsilon\Upsilon}^{-1} A_{\Upsilon\Gamma} + A_{B\Gamma}^T A_{BB}^{-1} A_{B\Gamma} & A_{\Upsilon\Gamma}^T & A_{B\Gamma}^T \\ A_{\Upsilon\Gamma} & A_{\Upsilon\Upsilon} & 0 \\ A_{B\Gamma} & 0 & A_{BB} \end{bmatrix} \tag{6.4.8}$$

for the stiff matrix of (6.4.2).

6.5 Numerical experiment

We present, in this section, some examples of application of Algorithm 6.2.1 using the Crouzeix–Raviart elements. The algorithm we used is based on the idea of determining two sequences of θ_1^n and θ_2^n converging as quickly as possible to the optimal values of the relaxation factors θ_1 and θ_2, which is given by (6.3.40). Although the constants σ and τ (given by (6.3.5) and (6.3.6)) are not known, we can build up a procedure which generates a sequence of nonconforming discrete harmonic functions on Ω_Υ and Ω_B with the same discrete values on Γ. This allows us to compute, at each iteration, two constants σ_n and τ_n as suggested by (6.3.5) and (6.3.6). Using these constants in (6.3.40) gives values of θ_1^n and θ_2^n to be used in our numerical scheme. We point out that the evaluation of θ_1^n and θ_2^n doesn't require the solution of any additional problem in our algorithm (for details cf.[41,65]). With this choice of the relaxation factors, our numerical experiment shows that the theoretical estimates are fully realized in practice.

The model problem is

$$\begin{cases} -\Delta u + u = f, & \text{in } \Omega = (0,1)^2, \\ u = 0, & \text{on } \partial\Omega. \end{cases} \tag{6.5.1}$$

Ω is divided into four subdomains:

$$\Omega_1 = (0, 0.75) \times (0.75, 1), \quad \Omega_2 = (0.75, 1)^2,$$

$$\Omega_3 = (0.75, 1) \times (0, 0.75), \quad \Omega_4 = (0, 0.75)^2.$$

There exists an internal cross point p; see Fig.6.5.1.

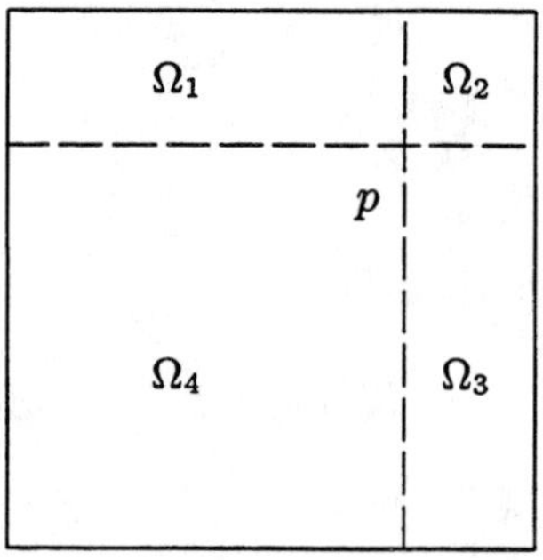

Fig.6.5.1

We triangulate Ω to get the fine mesh Ω_h such that each element $e \in \Omega_h$ is an isosceles right triangle with h as its diameter. When $h = 1/4$, there are, respectively, 12,4,12,36 elements and 22,8,22,60 interpolation points in $\overline{\Omega}_i, i = 1,2,3,4$. When $h = 1/8$, there are, respectively, 48,16,48,144 elements and 80,28,80,228 interpolation points in $\overline{\Omega}_i, i = 1,2,3,4$. The main experimental results are listed in Table 6.5.1 and Table 6.5.2. Here, n is the number of iterations, and $n = 0$ represents the initial situation. ε^n is the error function after n iterations. And,

$$E_n = A(\varepsilon^n, \varepsilon^n), \qquad \delta_n = \sqrt[n-1]{E_n/E_1},$$

$$M_n = \|\varepsilon^n\|_{L^\infty(\Omega)}, \qquad \rho_n = \sqrt[n-1]{M_n/M_1}.$$

Table 6.5.1: The errors when $h = 1/4$

n	0	5	10	15	20
E_n	$0.35 \cdot 10^7$	$0.45 \cdot 10^1$	$0.85 \cdot 10^{-3}$	$0.14 \cdot 10^{-5}$	$0.87 \cdot 10^{-8}$
M_n	$0.10 \cdot 10^4$	$0.83 \cdot 10^0$	$0.14 \cdot 10^{-1}$	$0.47 \cdot 10^{-3}$	$0.38 \cdot 10^{-4}$
δ_n		0.0338	0.0855	0.1303	0.1704
ρ_n		0.1699	0.2893	0.3532	0.4066

Table 6.5.2: The errors when $h = 1/8$

n	0	5	10	15	20
E_n	$0.12 \cdot 10^8$	$0.15 \cdot 10^6$	$0.56 \cdot 10^3$	$0.26 \cdot 10^1$	$0.21 \cdot 10^{-1}$
M_n	$0.10 \cdot 10^4$	$0.15 \cdot 10^3$	$0.88 \cdot 10^1$	$0.60 \cdot 10^0$	$0.41 \cdot 10^{-1}$
δ_n		0.3313	0.3296	0.3338	0.3360
ρ_n		0.6236	0.5909	0.5885	0.5879

Chapter 7

An Iterative Algorithm with Multi–subdomains for the Second Kind Nonconforming Elements ——in the General Subdivision Case

7.1 Introduction

The aim of the present chapter is to extend the trace averaging domain decomposition algorithm[9] to solve the elliptic problem

$$u \in H_0^1(\Omega): \ a(u,v) = (f,v), \ \forall\, v \in H_0^1(\Omega) \tag{7.1.1}$$

which is dicretized by the second kind nonconforming finite elements. In (7.1.1), $a(u,v) = \int_\Omega \left[\sum_{i,j=1}^{2} a_{ij}(x) \frac{\partial u}{\partial x_i} \frac{\partial v}{\partial x_j} + a_0(x) uv \right]$, $(f,v) = \int_\Omega fv$, $f \in H^{-1}(\Omega)$, $\Omega \subset \Re^2$ is a bounded polygonal open domain , $a_0(x) \geq \alpha > 0$, $a_0(x)$, $a_{ij}(x)$, $(i,j = 1,2)$ are piecewise smooth and bounded functions in Ω, (a_{ij}) is a symmetric, uniformly positive definite matrix in Ω.

Let $\Omega_h = \{e\}$ be a quasi–uniform mesh of Ω with h as its mesh parameter. Let V_h be the nonconforming finite element space of the second kind associated with Ω_h, which is defined in Chapter 1. The properties of V_h are well known. For any $v \in V_h$, v is continuous at the midpoint of each edge of e, $\forall\, e \in \Omega_h$, which is called the edge midpoint for conciseness. And, the optimal error estimates can be obtained for the Stokes problem, if V_h is used for the velocity field and piecewise constant elements for the pressure[27,53]. The nonconforming element discretization of (7.1.1) is

$$u_h \in V_h^0: \ A(u_h, v_h) = (f, v_h), \ \ \forall\, v_h \in V_h^0 \tag{7.1.2}$$

where $\qquad V_h^0 = \{v_h \in V_h: \ v_h(x) = 0, \ \forall \text{ edge midpoint } x \in \partial\Omega\},$

$$A(u,v) = \sum_{e \subset \Omega} \int_e \left[\sum_{i,j=1}^{2} a_{ij} \frac{\partial u}{\partial x_i} \frac{\partial v}{\partial x_j} + a_0 uv \right].$$

Decompose Ω into nonoverlap subdomains $\Omega_1, \Omega_2, \cdots, \Omega_N$, such that $\{\Omega_i\}_{i=1}^N$ is a quasi–uniform mesh of Ω with H as its mesh parameter and denoted Ω_H. Ω_i is either a quadrilateral or a triangle, referred to as a subdomain. Assume that Ω_h can be obtained by refining Ω_H, so that Ω_H and Ω_h form a two–level triangulation on Ω. Generally, $H \gg h$, therefore, Ω_H and Ω_h are called the coarse mesh and the fine mesh, respectively. To solve (7.1.2), Chapter 6 introduced an iterative substructuring method for the red–black subdivisions, while the present approach is applicable for general subdivisions. In each iteration of our algorithm, the Dirichlet subproblems set on Ω_i are solved simultaneously, which is then followed by the parallel solving of the Neumann subproblems posed on Ω_i. For it is unnecessary and in fact impossible to calculate the values at the internal cross points in the second kind nonconforming discrete cases, the internal cross points are not handled, which plus the simple way to interchange the information between the subdomains leads to the high parallelism of our algorithm. Another remarkable feature of the algorithm is that it is easy to implement. Based on Theorem 2.4.6, it is proved to be geometrically convergent. Virtually, it is equivalent to the simple iterative method applied to the preconditioned capacitance equation. Furthermore, the preconditioner implied in the iteration is easily invertible and can be used to accelerate the conjugate gradient method to solve (7.1.2).

The rest of this chapter is arranged as follows: Sect.7.2 describes the domain decomposition algorithm. Sect.7.3 gives the convergence analysis and matrix analysis. Finally, Sect.7.4 presents numerical experiments to show the efficiency of our algorithm.

7.2 Domain decomposition algorithm

$\bigcup_{i=1}^{N} \partial\Omega_i \backslash \partial\Omega$ is called the pseudo–boundary and denoted Γ. Let $\{\xi_j\}_{j=1}^{m}$ be the set of the edge midpoints on Γ (ordered in some way). Let $\{\phi_i^\Gamma\}$ be the set of the corresponding bases and m_i the number of the edge midpoints on $\partial\Omega_i \backslash \partial\Omega$. The discrete trace operator $r_0^h : V_h^0 \to \Re^m$ is defined as follows

$$\forall\, v \in V_h^0,\ r_0^h v \in \Re^m : \ \left(r_0^h v\right)(j) = v(\xi_j),\ j = 1, 2, \cdots, m.$$

Let $\beta_i = \{k : \Omega_k \in \Omega_H,\ \text{meas}(\partial\Omega_k \cap \partial\Omega_i) > 0\}$. For $j \in \beta_i$, the open common edge of Ω_i and Ω_j is labeled Γ_{ij}, and for $v \in V_h^0$, $\overline{v}_{ij}$, the piecewise linear continuous function on Γ_{ij}, satisfies

$$\overline{v}_{ij}(\xi_k) = v(\xi_k),\ \forall\, \xi_k \in \Gamma_{ij},\ \ \overline{v}_{ij}(\nu) = 0,\ \forall \text{ endpoint } \nu \text{ of } \Gamma_{ij}.$$

For $i = 1, 2, \cdots, N$, define a discrete trace averaging operator from Γ to $\partial\Omega_i \backslash \partial\Omega$ to be the matrix $R_i \in \Re^{m_i \times m}$ so that

$$(R_i)_{lj} = \begin{cases} \dfrac{1}{2}, & \text{if the } l\text{th edge midpoint on } \partial\Omega_i \backslash \partial\Omega \text{ is } \xi_j \\ 0, & \text{else} \end{cases}$$

where $(R_i)_{lj}$ represents the (l, j)th element of R_i. Finally we introduce the following notations:

$$V_h^i = \{v \in V_h^0 : \ v(x) = 0,\ \forall \text{ interpolation point } x \in \overline{\Omega} \backslash \overline{\Omega}_i\},$$

$$V_h^{i,0} = \{v \in V_h^0 : \ v(x) = 0, \ \forall \text{ interpolation point } x \in \overline{\Omega} \backslash \Omega_i\},$$

$$A_i(u,v) = \sum_{e \subset \Omega_i} \int_e \left[\sum_{k,j=1}^{2} a_{kj} \frac{\partial u}{\partial x_k} \frac{\partial v}{\partial x_j} + a_0 uv \right], \quad (f,v)_i = \int_{\Omega_i} fv.$$

Here, an interpolation point $x \in \overline{\Omega}$ is related to the definition of V_h.

Now we describe the trace averaging domain decomposition algorithm to solve the discrete problem (7.1.2).

Algorithm 7.2.1. Choose arbitrarily $\lambda^0 = (\lambda_1^0, \lambda_2^0, \cdots, \lambda_m^0)^T \in \Re^m$. Give a relaxation factor $\rho \in (0,1)$ and set n:=0.

Step 1. For $i = 1, 2, \cdots, N$, solve in parallel

$$\begin{cases} u_i^n \in V_h^i \\ A_i(u_i^n, \theta) = (f, \theta)_i , & \forall\, \theta \in V_h^{i,0} \\ u_i^n(\xi_j) = \lambda_j^n, & \forall\, \xi_j \in \partial\Omega_i \backslash \partial\Omega \end{cases}$$

Step 2. For $l = 1, 2, \cdots, m$, calculate in parallel

$$d_l^n = \frac{1}{2}\left[A_i(u_i^n, \phi_l^\Gamma) - (f, \phi_l^\Gamma)_i\right] + \frac{1}{2}\left[A_j(u_j^n, \phi_l^\Gamma) - (f, \phi_l^\Gamma)_j\right], \quad \text{if } \xi_l \in \partial\Omega_i \cap \partial\Omega_j$$

Step 3. For $i = 1, 2, \cdots, N$, solve in parallel

$$\begin{cases} \delta_i^n \in V_h^i \\ A_i(\delta_i^n, \theta) = 0 , & \forall\, \theta \in V_h^{i,0} \\ A_i(\delta_i^n, \phi_l^\Gamma) = d_l^n, & \forall\, \xi_l \in \partial\Omega_i \backslash \partial\Omega \end{cases}$$

Step 4. For $l = 1, 2, \cdots, m$, calculate in parallel

$$\lambda_l^{n+1} = \lambda_l^n - \frac{1}{2}\rho\Big(\delta_i^n(\xi_l) + \delta_j^n(\xi_l)\Big), \quad \text{if } \xi_l \in \partial\Omega_i \cap \partial\Omega_j$$

Set n:=n+1 and return to Step 1 until some reasonable stopping criterion is satisfied.

7.3 Theoretical analysis

Theorem 7.3.1. *Let u_h be the solution of (7.1.2), $u^n \in V_h^0$ the approximate solution obtained by Algorithm 7.2.1, where $u^n = u_i^n$ on Ω_i. $\varepsilon^n = u^n - u_h$ is the error. If $0 < \rho < \frac{2}{\sigma}$, then*

$$A(\varepsilon^{n+1}, \varepsilon^{n+1}) \le k(\rho) A(\varepsilon^n, \varepsilon^n), \tag{7.3.1}$$

where $k(\rho) = 1 - 2\rho + \sigma\rho^2 < 1$, *and σ is a postive constant so that*

$$1 \le \sigma \le c(1 + \ln\frac{H}{h})(1 + H^{-2} + \ln\frac{H}{h}). \tag{7.3.2}$$

Furthermore, $k(\rho_{opt}) = 1 - \frac{1}{\sigma} = \min\limits_{0<\rho<\frac{2}{\sigma}} k(\rho)$, *where* $\rho_{opt} = \frac{1}{\sigma}$.

Proof. For $i = 1, 2, \cdots, N$, define the linear operator $\Pi_i : V_h^0 \to V_h^i$ as follows

$$\forall\, v \in V_h^0,\ \Pi_i v \in V_h^i:\ \left(\Pi_i v\right)(x) = \begin{cases} 0, & \forall \text{ interpolation point } x \in \Omega_i \\ v(\xi_j), & \forall\, x = \xi_j \in \partial\Omega_i \backslash \partial\Omega \end{cases}$$

For the solution u_h of (7.1.2), let $u_h^i \in V_h^i$, which satisfies

$$u_h^i(x) = u_h(x),\ \forall \text{ interpolation point } x \in \overline{\Omega}_i.$$

Denote $\varepsilon_i^n = u_i^n - u_h^i$ and $\mu^n = \lambda^n - r_0^h u_h$. It is easy to see that

$$\begin{cases} \varepsilon_i^n \in V_h^i \\ A_i(\varepsilon_i^n, \theta) = 0, & \forall\, \theta \in V_h^{i,0} \\ \varepsilon_i^n(\xi_j) = (\mu^n)(j), & \forall\, \xi_j \in \partial\Omega_i \backslash \partial\Omega \end{cases} \tag{7.3.3}$$

$$\begin{cases} \delta_i^n \in V_h^i \\ A_i(\delta_i^n, \theta) = \dfrac{1}{2}\displaystyle\sum_{j=1}^{N} A_j(\varepsilon_j^n, \Pi_j\theta),\ \ \forall\, \theta \in V_h^i \end{cases} \tag{7.3.4}$$

$$\mu^{n+1} = \mu^n - \frac{1}{2}\rho \sum_{j=1}^{N} r_0^h \delta_j^n. \tag{7.3.5}$$

Let

$$\psi_i^n = \frac{1}{\rho}(\varepsilon_i^n - \varepsilon_i^{n+1}). \tag{7.3.6}$$

It follows from (7.3.3), (7.3.5) and (7.3.6) that

$$\Pi_i \psi_i^n = \Pi_i\Big(\frac{1}{2}\sum_{j=1}^{N} \delta_j^n\Big). \tag{7.3.7}$$

For $j \in \beta_i$, let $\omega_{ij}^n \in V_h^i$ satisfy

$$\begin{cases} A_i(\omega_{ij}^n, \theta) = 0, & \forall\, \theta \in V_h^{i,0} \\ \omega_{ij}^n(\xi_k) = \delta_j^n(\xi_k), & \forall\, \xi_k \in \Gamma_{ij} \\ \omega_{ij}^n(x) = 0, & \forall \text{ edge midpoint } x \in \partial\Omega_i \backslash \Gamma_{ij} \end{cases}$$

Then, Lemma 6.3.1 and Theorem 2.4.6 lead to

$$\begin{aligned} A_i(\omega_{ij}^n, \omega_{ij}^n) &\le c\|\overline{(\omega_{ij}^n)}_{ij}\|^2_{H_{00}^{\frac{1}{2}}(\Gamma_{ij})} = c\|\overline{(\delta_j^n)}_{ij}\|^2_{H_{00}^{\frac{1}{2}}(\Gamma_{ij})} \\ &\le c(1 + \ln \tfrac{H}{h})\Big\{(1 + \ln \tfrac{H}{h})|\delta_j^n|^2_{1,\Omega_j,h} + (1 + H^{-2})\|\delta_j^n\|^2_{L^2(\Omega_j)}\Big\} \\ &\le \sigma A_j(\delta_j^n, \delta_j^n), \end{aligned} \tag{7.3.8}$$

where σ is the constant defined by (7.3.2).

On the other hand, by (7.3.6), (7.3.3) and (7.3.5), we obtain

$$\psi_i^n = \frac{1}{2}\delta_i^n + \frac{1}{2}\sum_{j\in\beta_i}\omega_{ij}^n. \tag{7.3.9}$$

Furthermore, it follows from the triangle inequality and (7.3.8) that

$$A_i(\psi_i^n,\psi_i^n) \le c\Big\{A_i(\delta_i^n,\delta_i^n) + \sum_{j\in\beta_i} A_i(\omega_{ij}^n,\omega_{ij}^n)\Big\} \le \sigma\Big\{A_i(\delta_i^n,\delta_i^n) + \sum_{j\in\beta_i} A_j(\delta_j^n,\delta_j^n)\Big\}.$$

$$\sum_{i=1}^N A_i(\psi_i^n,\psi_i^n) \le \sigma\sum_{i=1}^N A_i(\delta_i^n,\delta_i^n). \tag{7.3.10}$$

By taking $\theta = \varepsilon_i^n$ in (7.3.4), summing up with i, and applying (7.3.3) , we have

$$\begin{aligned}\sum_{i=1}^N A_i(\delta_i^n,\varepsilon_i^n) &= \frac{1}{2}\sum_{i=1}^N\sum_{j=1}^N A_j(\varepsilon_j^n,\Pi_j\varepsilon_i^n)\\ &= \sum_{j=1}^N A_j\Big(\varepsilon_j^n,\Pi_j(\frac{1}{2}\sum_{i=1}^n\varepsilon_i^n)\Big) = \sum_{j=1}^n A_j(\varepsilon_j^n,\varepsilon_j^n),\end{aligned}$$

$$\sum_{i=1}^N A_i(\varepsilon_i^n,\varepsilon_i^n) = \sum_{i=1}^N A_i(\delta_i^n,\varepsilon_i^n) \le \Big(\sum_{i=1}^N A_i(\varepsilon_i^n,\varepsilon_i^n)\Big)^{\frac{1}{2}}\Big(\sum_{i=1}^N A_i(\delta_i^n,\delta_i^n)\Big)^{\frac{1}{2}},$$

which implies

$$\sum_{i=1}^N A_i(\varepsilon_i^n,\varepsilon_i^n) \le \sum_{i=1}^N A_i(\delta_i^n,\delta_i^n). \tag{7.3.11}$$

It follows from the substitution of θ in (7.3.3) with $\psi_i^n - \Pi_i\psi_i^n$ that

$$\sum_{i=1}^N A_i(\varepsilon_i^n,\psi_i^n) = \sum_{i=1}^N A_i(\varepsilon_i^n,\Pi_i\psi_i^n).$$

Furthermore, by taking $\theta = \delta_i^n$ in (7.3.4) and using (7.3.7) and (7.3.3), we get

$$\begin{aligned}\sum_{i=1}^N A_i(\delta_i^n,\delta_i^n) &= \frac{1}{2}\sum_{i=1}^N\sum_{j=1}^N A_j(\varepsilon_j^n,\Pi_j\delta_i^n)\\ &= \sum_{j=1}^N A_j\Big(\varepsilon_j^n,\Pi_j(\frac{1}{2}\sum_{i=1}^N\delta_i^n)\Big) = \sum_{j=1}^N A_j(\varepsilon_j^n,\psi_j^n).\end{aligned} \tag{7.3.12}$$

If $0 < \rho < \frac{2}{\sigma}$, then (7.3.12), 3.10) and (7.3.11) lead to

$$\begin{aligned}\sum_{i=1}^{N} A_i(\varepsilon_i^{n+1}, \varepsilon_i^{n+1}) &= \sum_{i=1}^{N} A_i(\varepsilon_i^n - \rho\psi_i^n, \varepsilon_i^n - \rho\psi_i^n) \\ &= \sum_{i=1}^{N} A_i(\varepsilon_i^n, \varepsilon_i^n) - 2\rho \sum_{i=1}^{N} A_i(\delta_i^n, \delta_i^n) + \rho^2 \sum_{i=1}^{N} A_i(\psi_i^n, \psi_i^n) \\ &\leq \sum_{i=1}^{N} A_i(\varepsilon_i^n, \varepsilon_i^n) - (2\rho - \sigma\rho^2) \sum_{i=1}^{N} A_i(\delta_i^n, \delta_i^n) \\ &\leq (1 - 2\rho + \sigma\rho^2) \sum_{i=1}^{N} A_i(\varepsilon_i^n, \varepsilon_i^n),\end{aligned}$$

which completes the proof of (7.3.1), and hence the theorem. □

Theorem 7.3.1 indicates that if $0 < \rho < \frac{2}{\sigma}$, then Algorithm 7.2.1 converges geometrically with the convergence factor $k(\rho) = 1 - 2\rho + \sigma\rho^2$, dependent on H, h. Now, we analyse Algorithm 7.2.1 from the algebraic viewpoint. Let $\{\phi_i\}$ be the set of the bases of V_h^0. ϕ_j^I represents the basis of V_h^0 in $\bigcup_{k=1}^{N} \Omega_k$. Suppose

$$u_h = \sum_i u_i^I \phi_i^I + \sum_{j=1}^{m} u_j^\Gamma \phi_j^\Gamma$$

is the solution of (7.1.2). Then $u_j^\Gamma = u_h(\xi_j)$, and (7.1.2) can be written as

$$KU = \begin{bmatrix} K_{II} & K_{I\Gamma} \\ K_{I\Gamma}^T & K_{\Gamma\Gamma} \end{bmatrix} \begin{bmatrix} U_I \\ U_\Gamma \end{bmatrix} = \begin{bmatrix} F_I \\ F_\Gamma \end{bmatrix}, \tag{7.3.13}$$

where

$$U_I = (u_i^I), \quad U_\Gamma = (u_j^\Gamma), \quad F_I = \big((f, \phi_i^I)\big), \quad F_\Gamma = \big((f, \phi_j^\Gamma)\big),$$

$$K = \big(A(\phi_i, \phi_j)\big), \quad K_{II} = \big(A(\phi_i^I, \phi_j^I)\big), \quad K_{I\Gamma} = \big(A(\phi_i^I, \phi_j^\Gamma)\big), \quad K_{\Gamma\Gamma} = \big(A(\phi_i^\Gamma, \phi_j^\Gamma)\big).$$

With the block Gaussian elimination, we obtain

$$SU_\Gamma = \tilde{F}_\Gamma\,, \tag{7.3.14}$$

where $\tilde{F}_\Gamma = F_\Gamma - K_{I\Gamma}^T K_{II}^{-1} F_I$, $S = K_{\Gamma\Gamma} - K_{I\Gamma}^T K_{II}^{-1} K_{I\Gamma}$ is the Schur complement or the capacitance matrix of Ω with regard to Γ. In the same way ,we can define the capacitance matrix of Ω_i with regard to $\partial\Omega_i \backslash \partial\Omega$, which is labeled S_i. It is easy to see that Algorithm 7.2.1 is essentially equivalent to the following iterative method for (7.3.14):

$$\lambda^{n+1} = \lambda^n - \rho \sum_{i=1}^{N} R_i^T S_i^{-1} R_i S \lambda^n + \rho \sum_{i=1}^{N} R_i^T S_i^{-1} R_i \tilde{F}_\Gamma. \tag{7.3.15}$$

Obviously, (7.3.15) can be viewed as the simple iterative method for (7.3.14) preconditioned by

$$Q = \Big(\rho \sum_{i=1}^{N} R_i^T S_i^{-1} R_i\Big)^{-1}. \tag{7.3.16}$$

Note that, Q is an efficient preconditioner of the capacitance matrix S, because it can be inversed easily in parallel and the condition number of $Q^{-1}S$ is bounded by $O\Big((1+\ln\frac{H}{h})(1+H^{-2}+\ln\frac{H}{h})\Big)$, which is implied by the following theorem.

Theorem 7.3.2. *Let S and Q be those in (7.3.14) and (7.3.16). Then* $\forall\ \lambda \in \Re^m\backslash\{0\}$

$$\rho \leq \frac{(SQ^{-1}S\lambda,\lambda)}{(S\lambda,\lambda)} \leq c\rho(1+\ln\frac{H}{h})(1+H^{-2}+\ln\frac{H}{h}). \tag{7.3.17}$$

Proof. Let ε_i, δ_i be generated by Algorithm 7.2.1 with $f=0$ and $\lambda^n = \lambda$. A more iteration yields $\overline{\varepsilon}_i, \overline{\delta}_i$. Denote $\psi_i = \frac{1}{\rho}(\varepsilon_i - \overline{\varepsilon}_i)$. It is easy to see that

$$(S\lambda,\lambda) = \sum_{i=1}^{N} A_i(\varepsilon_i,\varepsilon_i), \quad (SQ^{-1}S\lambda,\lambda) = \rho\sum_{i=1}^{N} A_i(\varepsilon_i,\psi_i).$$

Thus, by (7.3.12) and (7.3.11), we get the left hand side of (7.3.17).

On the other hand, it follows from (7.3.12) and (7.3.10) that

$$\begin{aligned}\sum_{i=1}^{N} A_i(\delta_i,\delta_i) = \sum_{i=1}^{N} A_i(\varepsilon_i,\psi_i) &\leq \Big(\sum_{i=1}^{N} A_i(\varepsilon_i,\varepsilon_i)\Big)^{\frac{1}{2}}\Big(\sum_{i=1}^{N} A_i(\psi_i,\psi_i)\Big)^{\frac{1}{2}}\\ &\leq \Big(\sum_{i=1}^{N} A_i(\varepsilon_i,\varepsilon_i)\Big)^{\frac{1}{2}}\Big(\sigma\sum_{i=1}^{N} A_i(\delta_i,\delta_i)\Big)^{\frac{1}{2}},\end{aligned}$$

which implies the right hand side of (7.3.17) holds. Hence the proof of the theorem is finished. □

Remark 7.3.1. It follows from the properties of Q defined by (7.3.16) that

$$\begin{bmatrix} K_{II} & K_{I\Gamma} \\ K_{I\Gamma}^T & Q + K_{I\Gamma}^T K_{II}^{-1} K_{I\Gamma} \end{bmatrix}$$

is an efficient preconditioner of the stiff matrix in (7.3.13).

7.4 Numerical experiments

We consider the following model problem on the unit square Ω:

$$\begin{cases} -\Delta u + u = f, & \text{in } \Omega \\ u = 0, & \text{on } \partial\Omega \end{cases} \tag{7.4.1}$$

which is discretized by the Crouzeix–Raviart nonconforming elements.

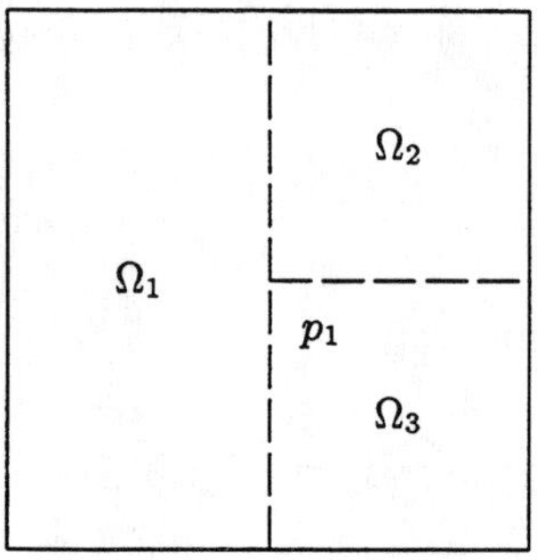

Fig.7.4.1

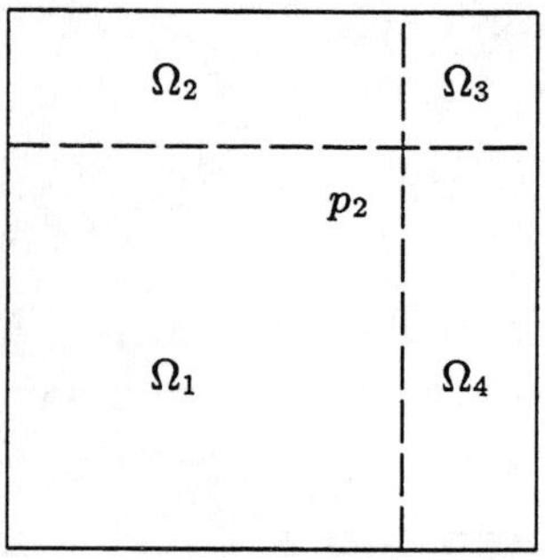

Fig.7.4.2

Example 7.4.1. Ω is divided into three nonoverlap subdomains:

$$\Omega_1 = (0, 0.5) \times (0, 1), \quad \Omega_2 = (0.5, 1) \times (0.5, 1), \quad \Omega_3 = (0.5, 1) \times (0, 0.5),$$

which share an internal cross point p_1, but do not satisfy the red–black subdivision; see Fig.7.4.1. We triangulate Ω to get the fine mesh Ω_h so that each element $e \in \Omega_h$ is an isosceles right triangle with h as its diameter. When $h = 1/4$, there are 32, 16, and 16 elements, 54, 28, and 28 nodes in Ω_1, Ω_2, Ω_3, respectively. When $h = 1/8$, there are 128, 64, and 64 elements, 204, 104, and 104 nodes in Ω_1, Ω_2, Ω_3, respectively. In Table 7.4.1, n represents the number of iterations, ε^n the error function as that in Theorem 7.3.1, ρ the relaxation factor, and $\delta_n = \sqrt[n-1]{E_n/E_1}$ the average reduction factor, where $E_n = A(\varepsilon^n, \varepsilon^n)$.

Table 7.4.1 illustrates that Algorithm 7.2.1 converges faster when $h = 1/4$ than $h = 1/8$, which is in accordance with our theoretical analysis. We also observe that the optimal relaxation factor ρ_{opt} depends on h.

Example 7.4.2. We decompose Ω into four subdomains:

$$\Omega_1 = (0, 0.75) \times (0, 0.75), \quad \Omega_2 = (0, 0.75) \times (0.75, 1),$$

$$\Omega_3 = (0.75, 1) \times (0.75, 1), \quad \Omega_4 = (0.75, 1) \times (0, 0.75),$$

which share an internal cross point p_2 and satisfy the red–black subdivision; see Fig.7.4.2. The fine mesh Ω_h is the same as that in Example 7.4.1. In Ω_i $(i = 1, 2, 3, 4)$ there are, respectively, 36, 12, 4 and 12 elements, 60, 22, 8, and 22 nodes for $h = 1/4$, 144, 48, 16, and 48 elements, 228, 80, 28, and 80 nodes for $h = 1/8$. The Numerical results are listed in Table 7.4.2, from which we see that for the red–black subdivision, Algorithm 7.2.1 can perform adaptively (for details, cf.[41]) to avoid the blindness of choosing ρ, therefore, it converges much faster.

Table 7.4.1: δ_n vs. ρ, h for Example 7.4.1

n		2	6	10	20	30
$h = 1/4$	$\rho = 0.45$	0.320	0.320	0.289	0.313	0.328
	$\rho = 0.40$	0.339	0.294	0.362	0.381	0.375
	$\rho = 0.35$	0.398	0.296	0.413	0.423	0.425
	$\rho = 0.30$	0.395	0.395	0.399	0.490	0.494
	$\rho = 0.20$	0.507	0.491	0.545	0.598	0.623
$h = 1/8$	$\rho = 0.45$	0.566	0.575	0.689	0.790	0.824
	$\rho = 0.40$	0.479	0.369	0.340	0.387	0.396
	$\rho = 0.35$	0.425	0.419	0.426	0.437	0.447
	$\rho = 0.30$	0.405	0.499	0.483	0.500	0.508
	$\rho = 0.20$	0.568	0.603	0.639	0.645	0.647

Table 7.4.2: E_n, δ_n vs. h for Example 7.4.2

n		5	10	20	30
$h = 1/4$	E_n	$0.14 \cdot 10^5$	$0.15 \cdot 10^2$	$0.16 \cdot 10^{-4}$	$0.17 \cdot 10^{-10}$
	δ_n	0.253	0.253	0.253	0.253
	ρ_n	0.495	0.495	0.496	0.496
$h = 1/8$	E_n	$0.18 \cdot 10^5$	$0.20 \cdot 10^2$	$0.23 \cdot 10^{-4}$	$0.27 \cdot 10^{-10}$
	δ_n	0.254	0.254	0.255	0.255
	ρ_n	0.494	0.495	0.495	0.495

Chapter 8

Preconditioners with Multi–subdomains for 2nd Kind Nonconforming Elements by Substructuring

8.1 Introduction and notations

Consider the variational problem

$$u \in H_0^1(\Omega): \ a(u,v) = (f,v), \ \forall \ v \in H_0^1(\Omega) \tag{8.1.1}$$

where $a(u,v) = \int_\Omega \Big[\sum_{i,j=1}^{2} a_{ij}(x)\frac{\partial u}{\partial x_i}\frac{\partial v}{\partial x_j} + a_0(x)uv\Big]$, $(f,v) = \int_\Omega fv$, $f \in H^{-1}(\Omega)$, $\Omega \subset \Re^2$ is a bounded polygonal open domain, $a_0(x)$, $a_{ij}(x)$, $(i,j = 1,2)$ are piecewise smooth and bounded functions in Ω, $a_0(x) \geq 0$ in Ω and (a_{ij}) is a symmetric, uniformly positive definite matrix in Ω. Let $\Omega_h = \{e\}$ be a quasi–uniform mesh of Ω with h as its mesh parameter. Let V_h be the second kind nonconforming finite element space on Ω_h. The nonconforming element discretization of (8.1.1) is

$$u_h \in V_h^0: \ A(u_h,v) = (f,v), \ \ \forall \ v \in V_h^0 \tag{8.1.2}$$

where

$$V_h^0 = \{v \in V_h: \ v(x) = 0, \ \forall \text{ interpolation point } x \in \partial\Omega\},$$

$$A(w,v) = \sum_{e\subset\Omega}\int_e \Big[\sum_{i,j=1}^{2} a_{ij}\frac{\partial w}{\partial x_i}\frac{\partial v}{\partial x_j} + a_0wv\Big].$$

Let $\{\Omega_k\}_{k=1}^N \triangleq \Omega_H$ be a nonoverlap subdivision of Ω, which is, in fact, a quasi–uniform mesh of Ω with H as its mesh parameter. Ω_i is either a quadrilateral or a triangle, referred to as the subdomain. Besides, the common boundary point of more than two subdomains that lies inside Ω is called the internal cross point. Assume that Ω_h can be obtained by refining Ω_H, so that Ω_H and Ω_h form a two–level triangulation on Ω. Generally, $H \gg h$, therefore, Ω_H and Ω_h are called the coarse mesh and the fine mesh respectively. In the conforming discrete cases, Bramble[10−15], et al., have developed many preconditioners which are easily inversed in parallel and can decrease the condition number of the discrete system from $O(h^{-2})$ to $O\Big((1+\ln\frac{H}{h})^k\Big)$, $(k = 2,3)$. In the preconditioning algorithms, the coarse mesh problems must be solved to separate the internal cross points from other mesh nodes. But when considering (8.1.2), we find that it is unnecessary and in fact impossible to obtain the values of u_h at the internal cross points. With this in mind, we develop some simpler preconditioners for (8.1.2) in this chapter which can be inverted in high parallel. With the essential estimates in Chapter 2, we show that the condition number of the preconditioned system is $O\Big((1+\ln\frac{H}{h})(1+H^{-2}+\ln\frac{H}{h})\Big)$. Thus, they are very efficient; see [51] also.

In order to describe the preconditioners in Sect.8.2, we introduce some notations. Let $\{\xi_j\}_{j=1}^m$ be the set of the vertices of $\Omega_1, \Omega_2, \cdots, \Omega_N$ (ordered in some way). Let $\Gamma_{ij} \not\subset \partial\Omega$ be the open edge with ξ_i, ξ_j as its endpoints which is a boundary edge of some subdomain Ω_k. Denote

$$V_h^k = \{v \in V_h^0 :\ v(x) = 0,\ \forall \text{ interpolation point } x \in \Omega\backslash\overline{\Omega}_k\},$$

$$V_h^{k,0} = \{v \in V_h^0 :\ v(x) = 0,\ \forall \text{ interpolation point } x \in \Omega\backslash\Omega_k\},$$

$$A_k(w,v) = \sum_{e\subset\Omega_k} \int_e \Big[\sum_{i,j=1}^{2} a_{ij}\frac{\partial w}{\partial x_i}\frac{\partial v}{\partial x_j} + a_0 wv\Big],$$

$$\beta_k = \{(i,j) :\ \Gamma_{ij} \subset \partial\Omega_k,\ 1 \le i,j \le m\}.$$

If $v \in V_h$ satisfies

$$A_k(v,w) = 0,\ \forall\ w \in V_h^{k,0},$$

we call v *the discrete harmonic function* on Ω_k. For each $v \in V_h$, $\overline{v}_{ij}$ denotes the piecewise linear continuous function on $\bigcup_{k=1}^{N} \partial\Omega_k$, which satisfies

$$\overline{v}_{ij} = 0, \text{ on } \bigcup_{k=1}^{N} \partial\Omega_k\backslash\Gamma_{ij},\quad \overline{v}_{ij}(x) = v(x),\ \forall \text{ interpolation point } x \in \Gamma_{ij}$$

and let $\hat{v}_{ij} \in V_h$ be the discrete harmonic function on $\Omega_1, \Omega_2, \cdots, \Omega_N$, s.t.

$$\hat{v}_{ij}(x) = \overline{v}_{ij}(x),\ \forall \text{ interpolation point } x \in \bigcup_{k=1}^{N} \partial\Omega_k.$$

In addition, v_{ij} denotes the vector of the values of v at the interpolation points on Γ_{ij}. Let $S_0^h(\Gamma_{ij}) \subset H_0^1(\Gamma_{ij})$ be the piecewise linear continuous finite element space. We note that the operator $l_{\Gamma_{ij}}$ defined on $S_0^h(\Gamma_{ij})$ by

$$\langle l_{\Gamma_{ij}} v, w\rangle_{\Gamma_{ij}} = \int_{\Gamma_{ij}} \frac{dv}{ds}\frac{dw}{ds} ds,\ \forall\ w \in S_0^h(\Gamma_{ij})$$

is a finite–dimensional approximation to the Laplace operator $-\dfrac{d^2}{ds^2}$. Here, s is the arc length along Γ_{ij}. Let $l_{\Gamma_{ij}}^{\frac{1}{2}}$ be the positive square root of the symmetric, positive definite matrix $l_{\Gamma_{ij}}$. It follows from [12,60] that

$$c\langle l_{\Gamma_{ij}}^{\frac{1}{2}} w, w\rangle_{\Gamma_{ij}} \le \|w\|^2_{H_{00}^{\frac{1}{2}}(\Gamma_{ij})} \le C\langle l_{\Gamma_{ij}}^{\frac{1}{2}} w, w\rangle_{\Gamma_{ij}},\ \forall\ w \in S_0^h(\Gamma_{ij}) \qquad (8.1.3)$$

8.2 Preconditioners

By the Rayleigh theorem, the inverse inequality[26] and Theorem 2.4.1, we can show that the condition number of the nonconforming finite element discrete problem (8.1.2) is $O(h^{-2})$. When h is small enough, (8.1.2) is rather ill–conditioned. The preconditioned conjugate gradient method (PCG) is preferable to solve (8.1.2). For this purpose, as in [12], another symmetric bilinear form $Q(\cdot,\cdot)$ must be constructed to satisfy

(1) For a given function g, the solution of the following problem is easier in parallel to obtain than that of (8.1.2)

$$v \in V_h^0: \ Q(v,w) = (g,w), \ \forall\, w \in V_h^0 \tag{8.2.1}$$

(2) There exist two positive constants λ_0 and λ_1; $\dfrac{\lambda_1}{\lambda_0}$ is as small as possible, s.t.

$$\lambda_0 Q(w,w) \le A(w,w) \le \lambda_1 Q(w,w), \ \forall\, w \in V_h^0 \tag{8.2.2}$$

If $Q(\cdot,\cdot)$ satisfies (8.2.1) and (8.2.2), we call $Q(\cdot,\cdot)$ the preconditioner of (8.1.2). As in [12], $Q(\cdot,\cdot)$ is constructed by the decomposition of the domain and the function. To this end, we decompose $w \in V_h^0$ as $w = w_p + w_H$, s.t.

$$\begin{cases} w_p \in V_h^{1,0} \oplus V_h^{2,0} \oplus \cdots \oplus V_h^{N,0} \\ A(w_p, v) = A(w,v), \ \forall\, v \in V_h^{1,0} \oplus V_h^{2,0} \oplus \cdots \oplus V_h^{N,0} \end{cases} \tag{8.2.3}$$

$$\begin{cases} w_H \in V_h^0 \\ A(w_H, v) = 0, \quad \forall\, v \in V_h^{1,0} \oplus V_h^{2,0} \oplus \cdots \oplus V_h^{N,0} \\ w_H(x) = w(x), \ \ \forall \text{ interpolation point } x \in \bigcup\limits_{k=1}^{N} \partial\Omega_k \end{cases} \tag{8.2.4}$$

Obviously, w_H is discrete harmonic on $\Omega_1, \Omega_2, \cdots, \Omega_N$ and we have

$$A(w,w) = A(w_p, w_p) + A(w_H, w_H). \tag{8.2.5}$$

Note that w_p can be obtained by solving previously independent subproblems (8.2.3). Thus, the key to constructing the preconditioner $Q(\cdot,\cdot)$ is to find some proper form to replace $A(w_H, w_H) = \sum\limits_{k=1}^{N} A_k(w_H, w_H)$ in (8.2.5).

Theorem 8.2.1. *The bilinear form*

$$Q(v,w) = A(v_p, w_p) + \sum_{\Gamma_{ij}} \langle l_{\Gamma_{ij}}^{\frac{1}{2}} \overline{(v_H)}_{ij}, \overline{(w_H)}_{ij} \rangle_{\Gamma_{ij}}, \tag{8.2.6}$$

satisfies that $\forall\, w \in V_h^0$

$$c\Big\{(1 + \ln\frac{H}{h})(1 + H^{-2} + \ln\frac{H}{h})\Big\}^{-1} Q(w,w) \le A(w,w) \le C Q(w,w). \tag{8.2.7}$$

Proof. The right hand side of (8.2.7) follows from (8.2.5), Theorem 2.4.3 and (8.1.3). In addition, Theorem 2.4.6 and Theorem 2.3.1 imply that

$$\begin{aligned}
\sum_{\Gamma_{ij}} \langle l_{\Gamma_{ij}}^{\frac{1}{2}} \overline{(w_H)}_{ij}, \overline{(w_H)}_{ij} \rangle_{\Gamma_{ij}} &\le c \sum_{\Gamma_{ij}} \| \overline{(w_H)}_{ij} \|^2_{H_{00}^{\frac{1}{2}}(\Gamma_{ij})} \\
&\le c \sum_{k=1}^{N} (1 + \ln \frac{H}{h}) \Big\{ (1 + \ln \frac{H}{h}) |w_H|^2_{1,\Omega_k,h} + (1 + H^{-2}) \|w_H\|^2_{L^2(\Omega_k)} \Big\} \\
&\le c(1 + \ln \frac{H}{h}) \Big\{ (1 + \ln \frac{H}{h}) |w_H|^2_{1,\Omega,h} + (1 + H^{-2}) \|w_H\|^2_{L^2(\Omega)} \Big\} \\
&\le c(1 + \ln \frac{H}{h})(1 + H^{-2} + \ln \frac{H}{h}) |w_H|^2_{1,\Omega,h} \\
&\le c(1 + \ln \frac{H}{h})(1 + H^{-2} + \ln \frac{H}{h}) A(w_H, w_H),
\end{aligned}$$

so by (8.2.5), the left hand side of (8.2.7) holds. □

When using PCG with $Q(\cdot,\cdot)$ as preconditioner to obtain the solution of (8.1.2), we must solve (8.2.1) at each iteration. The steps to solve (8.2.1) is summarized as follows:

Algorithm 8.2.1.

Step 1. Find v_p by solving in parallel the homogeneous Dirichlet subproblems

$$v_p \in V_h^{1,0} \oplus V_h^{2,0} \oplus \cdots \oplus V_h^{N,0} : \; A_k(v_p, w) = (g, w), \; \forall \, w \in V_h^{k,0} \tag{8.2.8}$$

Step 2. Find $\overline{(v_H)}_{ij}$ by solving in parallel one–dimensional equations on each Γ_{ij}

$$\langle l_{\Gamma_{ij}}^{\frac{1}{2}} \overline{(v_H)}_{ij}, \overline{(w_H)}_{ij} \rangle_{\Gamma_{ij}} = (g, w) - A(v_p, w), \quad \forall \, w \in \Phi^{ij} \tag{8.2.9}$$

where $\Phi^{ij} \triangleq \text{span}\{\phi_n^{\Gamma_{ij}}\}$ and $\phi_n^{\Gamma_{ij}}$ represents a typical basis of V_h^0 concerning an interpolation point on Γ_{ij}.

Step 3. Find v_H by solving (8.2.4) in parallel, i.e., the discrete harmonic extension to each Ω_k with the values of v_H at the interpolation points on $\bigcup_{k=1}^{N} \partial\Omega_k \backslash \partial\Omega$ obtained by Step 2.

Step 4. Set $v = v_p + v_H$.

Theorem 8.2.1 shows that $Q(\cdot,\cdot)$ can decrease the condition number of the discrete system (8.1.2) from $O(h^{-2})$ to $O\big((1+\ln \frac{H}{h})(1+H^{-2}+\ln \frac{H}{h})\big)$, while Algorithm 8.2.1 indicates that (8.2.1) can be solved in high parallel. We note that Algorithm 8.2.1 is irrelevant to the internal cross points, which is the originality of the present approach. Therefore, $Q(\cdot,\cdot)$ is an efficient preconditioner for (8.1.2).

Suppose $\Gamma_{ij} \subset \partial\Omega_k$. Let $\{\phi_l^k\}$ be the set of bases of $V_h^{k,0}$. Denote

$$K = \big(A_k(\phi_l^k, \phi_n^k)\big), \qquad K_\Gamma = \big(A_k(\phi_l^k, \phi_n^{\Gamma_{ij}})\big),$$
$$K_{\Gamma\Gamma} = \big(A_k(\phi_l^{\Gamma_{ij}}, \phi_n^{\Gamma_{ij}})\big), \quad S_{ij}^k = K_{\Gamma\Gamma} - K_\Gamma^T K^{-1} K_\Gamma.$$

We call S_{ij}^k the capacitance matrix of Ω_k concerning Γ_{ij}. From (8.1.3) and Theorem 2.4.3, it follows that $\langle l_{\Gamma_{ij}}^{\frac{1}{2}} \overline{(w_H)_{ij}}, \overline{(w_H)_{ij}} \rangle_{\Gamma_{ij}}$ is equivalent to $A_k(\widehat{(w_H)}_{ij}, \widehat{(w_H)}_{ij})$. Thus, we have

Corollary 8.2.2. *If the second term of the right hand side of (8.2.6) is replaced by*

$$\sum_{\Gamma_{ij}} A_k(\widehat{(v_H)}_{ij}, \widehat{(w_H)}_{ij}) = \sum_{\Gamma_{ij}} (v_H)_{ij}^T S_{ij}^k (w_H)_{ij}, \tag{8.2.10}$$

then (8.2.7) is still true. In (8.2.10), Ω_k is one subdomain s.t. $(i,j) \in \beta_k$.

Remark 8.2.1. If the second term of the right hand side of (8.2.6) is replaced by (8.2.10), only Step 2 of Algorithm 8.2.1 is different. In this case, Step 2 is: Find $\widehat{(v_H)}_{ij}$ by solving simultaneously the discrete harmonic subproblems with the homogeneous Dirichlet condition on $\partial\Omega_k \backslash \Gamma_{ij}$ and the Neumann condition on Γ_{ij}.

Remark 8.2.2. In (3.5.2), (3.5.3) and (3.5.4), three matrices D, G and J are introduced, which are all inverted easily by FFT and spectrally equivalent to the capacitance matrix. For Γ_{ij}, they are denoted D_{ij}, G_{ij} and J_{ij} respectively. Therefore, if the second term of the right hand side of (8.2.6) is replaced by

$$\sum_{\Gamma_{ij}} (v_H)_{ij}^T D_{ij} (w_H)_{ij}, \quad \sum_{\Gamma_{ij}} (v_H)_{ij}^T G_{ij} (w_H)_{ij}, \text{ or } \sum_{\Gamma_{ij}} (v_H)_{ij}^T J_{ij} (w_H)_{ij},$$

the only difference is that Step 2 of Algorithm 8.2.1 can be performed by FFT.

To end this section, let's give the matrix analysis of the following preconditioner

$$\tilde{Q}(v, w) = A(v_p, w_p) + \sum_{\Gamma_{ij}} A_k(\widehat{(v_H)}_{ij}, \widehat{(w_H)}_{ij}). \tag{8.2.11}$$

In (8.2.11), Ω_k is one subdomain s.t. $(i,j) \in \beta_k$.

Suppose that the interpolation points of V_h^0 are labelled in the order: first those in $\Omega_1, \Omega_2, \cdots, \Omega_N$, then those on Γ_{ij}. Denote

$$\{\phi_j^I\} = \bigcup_{k=1}^{N} \{\phi_i^k\}, \qquad \{\phi_l^\Gamma\} = \bigcup_{\Gamma_{ij}} \{\phi_n^{\Gamma_{ij}}\},$$

$$A_{II} = \Big(A(\phi_l^I, \phi_n^I)\Big), \quad A_{I\Gamma} = \Big(A(\phi_l^I, \phi_n^\Gamma)\Big), \quad S = block - diag\Big(S_{ij}^k\Big),$$

where the matrix S is determined by all Γ_{ij}. Let $v = \sum_i \alpha_i^I \phi_i^I + \sum_j \alpha_j^\Gamma \phi_j^\Gamma$ be the solution of (8.2.1) with $\tilde{Q}(\cdot,\cdot)$ instead of $Q(\cdot,\cdot)$. In this case, (8.2.1) can be rewritten as

$$\begin{bmatrix} A_{II} & A_{I\Gamma} \\ A_{I\Gamma}^T & S + A_{I\Gamma}^T A_{II}^{-1} A_{I\Gamma} \end{bmatrix} \begin{bmatrix} v_I \\ v_\Gamma \end{bmatrix} = \begin{bmatrix} g_I \\ g_\Gamma \end{bmatrix}. \tag{8.2.12}$$

With the block Gaussian elimination, we can solve (8.2.12) by

Algorithm 8.2.2.

Step 1. Solve $v_I' = A_{II}^{-1} g_I$.

Step 2. Solve $v_\Gamma = S^{-1}(g_\Gamma - A_{I\Gamma}^T v_I')$.
Step 3. Solve $v_I'' = -A_{II}^{-1} A_{I\Gamma} v_\Gamma$.
Step 4. Set $v_I = v_I' + v_I''$.

Since A_{II} is a block–diagonal matrix, A_{II}^{-1} can be calculated simultaneously through the inverse of each diagonal block which is equivalent to the solving of the homogeneous Dirichlet subproblem on each Ω_k. Hence, Step 1 of Algorithm 8.2.2 is virtually Step 1 of Algorithm 8.2.1. In Step 2 of Algorithm 8.2.2, S^{-1} can also be calculated simultaneously through the inverse of each diagonal block which is equivalent to the solving of the discrete harmonic subproblem on some Ω_k with the Neumann condition on $\Gamma_{ij} \subset \partial\Omega_k$ and the homogeneous Dirichlet condition on $\partial\Omega_k \backslash \Gamma_{ij}$. Therefore, Step 2 of Algorithm 8.2.2 is different from Step 2 of Algorithm 8.2.1. Besides, it is not difficult to see that Step 3 and Step 4 of Algorithm 8.2.2 are the matrix representations of those of Algorithm 8.2.1 respectively.

We point out that the matrix analysis for other preconditioners can be carried out similarly.

8.3 Numerical experiments

We consider the following model problem on the unit square Ω :

$$\begin{cases} \Delta u = f, & \text{in } \Omega \\ u = 0, & \text{on } \partial\Omega \end{cases} \tag{8.3.1}$$

which is discretized by the Crouzeix–Raviart nonconforming elements. Ω is divided into four nonoverlap subdomains:

$$\Omega_1 = (0, 0.5) \times (0, 0.5), \quad \Omega_2 = (0, 0.5) \times (0.5, 1),$$

$$\Omega_3 = (0.5, 1) \times (0.5, 1), \quad \Omega_4 = (0.5, 1) \times (0, 0.5),$$

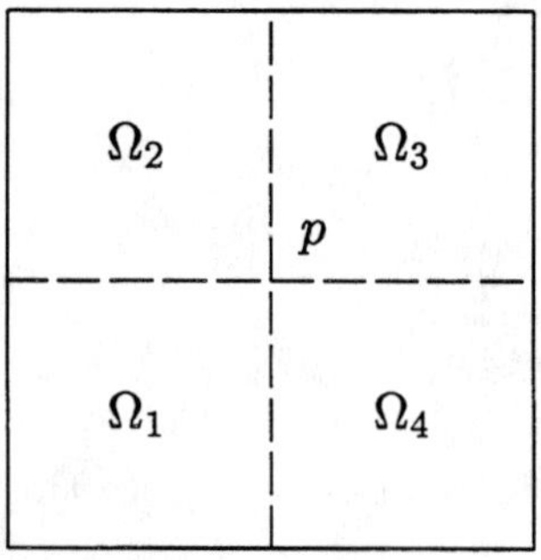

Fig.8.3.1

which share an internal cross point p. See Fig.8.3.1. We triangulate Ω to get the fine mesh Ω_h so that each $e \in \Omega_h$ is an isosceles right triangle with h as its diameter. When $h = 1/4$, there are 16 elements, 28 interpolation points in each Ω_k. When

$h = 1/8$, there are 64 elements, 104 interpolation points in each Ω_k. To solve (8.3.1), we apply the conjuate gradient method (CG) and PCG with $\tilde{Q}(\cdot,\cdot)$ in (8.2.11) as the preconditioner. The numerical results are listed in Table 8.3.1–Table 8.3.5 to compare CG and PCG with the same initial guess. Here, n represents the number of iterations, ε^n is the error function after n iterations, and

$$\|\varepsilon^n\|_A = \sqrt{A(\varepsilon^n, \varepsilon^n)}, \qquad \|\varepsilon^n\|_\infty = \max_{x\in\Omega} |\varepsilon^n(x)|,$$
$$\rho_n = \sqrt[n]{\|\varepsilon^n\|_A / \|\varepsilon^0\|_A}, \qquad \delta_n = \sqrt[n]{\|\varepsilon^n\|_\infty / \|\varepsilon^0\|_\infty}.$$

We observe from Table 8.3.1–Table 8.3.5 that the numerical results coincide with our theoretical analysis and it is remarkably effective to use $\tilde{Q}(\cdot,\cdot)$ in (8.2.11) to accelerate the convergence of solving (8.3.1) by CG.

Table 8.3.1: Applying CG in the case $h = 0.25$

n	0	5	10	15	20	25
$\|\varepsilon^n\|_A$	$0.32 \cdot 10^8$	$0.64 \cdot 10^5$	$0.26 \cdot 10^4$	$0.19 \cdot 10^3$	$0.77 \cdot 10^{-2}$	$0.15 \cdot 10^{-6}$
$\|\varepsilon^n\|_\infty$	$0.10 \cdot 10^4$	$0.85 \cdot 10^2$	$0.21 \cdot 10^2$	$0.40 \cdot 10^1$	$0.21 \cdot 10^{-1}$	$0.11 \cdot 10^{-3}$
ρ_n		0.2885	0.3900	0.4482	0.3305	0.2674
δ_n		0.6103	0.6789	0.6925	0.5838	0.5275

Table 8.3.2: Applying PCG in the case $h = 0.25$

n	0	1	2	3	4
$\|\varepsilon^n\|_A$	$0.32 \cdot 10^8$	$0.65 \cdot 10^1$	$0.27 \cdot 10^0$	$0.14 \cdot 10^{-6}$	$0.62 \cdot 10^{-7}$
$\|\varepsilon^n\|_\infty$	$0.10 \cdot 10^4$	$0.56 \cdot 10^0$	$0.25 \cdot 10^0$	$0.13 \cdot 10^{-3}$	$0.96 \cdot 10^{-4}$
ρ_n		0.0000002	0.0000916	0.0000162	0.0002101
δ_n		0.0005588	0.0156971	0.0050535	0.0175844

Table 8.3.3: Applying CG in the case $h = 0.125$

n	0	10	20	30	40	50
$\|\varepsilon^n\|_A$	$0.64 \cdot 10^8$	$0.80 \cdot 10^4$	$0.34 \cdot 10^3$	$0.34 \cdot 10^2$	$0.60 \cdot 10^{-1}$	$0.77 \cdot 10^{-5}$
$\|\varepsilon^n\|_\infty$	$0.10 \cdot 10^4$	$0.27 \cdot 10^2$	$0.80 \cdot 10^1$	$0.25 \cdot 10^1$	$0.43 \cdot 10^{-1}$	$0.61 \cdot 10^{-3}$
ρ_n		0.4070	0.5445	0.6180	0.5947	0.5516
δ_n		0.6975	0.7857	0.8193	0.7778	0.7511

Table 8.3.4: Applying PCG in the case $h = 0.125$

n	0	1	2	3	4	5
$\|\varepsilon^n\|_A$	$0.64 \cdot 10^8$	$0.11 \cdot 10^2$	$0.73 \cdot 10^0$	$0.17 \cdot 10^0$	$0.11 \cdot 10^{-3}$	$0.92 \cdot 10^{-7}$
$\|\varepsilon^n\|_\infty$	$0.10 \cdot 10^4$	$0.79 \cdot 10^1$	$0.49 \cdot 10^0$	$0.18 \cdot 10^0$	$0.22 \cdot 10^{-2}$	$0.87 \cdot 10^{-4}$
ρ_n		0.0000002	0.0001068	0.0013793	0.0011340	0.0010744
δ_n		0.0007889	0.0221873	0.0561342	0.0386572	0.0387534

Table 8.3.5: Iterations & CPU time when $\|\varepsilon^n\|_A \leq 10^{-7}$

		iterations	CPU time (second)
CG method	$h = 0.250$	27	7
	$h = 0.125$	56	21
PCG method	$h = 0.250$	4	3
	$h = 0.125$	5	12

Chapter 9

Overlap Domain Decomposition Methods

9.1 For nonselfadjoint problems discretized by Crouzeix–Raviart elements

9.1.1 Introduction

With the development of parallel computers, the idea of domain decomposition[76] has been paid more and more attention in the area of large scale scientific computing. Being one of the most active branches of current computational mathematics, domain decomposition methods consist of parallel computing, preconditioning technique, multigrid, multilevel and fast algorithm[36,38]. Overlap domain decomposition method, also known as the Schwarz alternating method, is the most original and fundamental domain decomposition. Based on the maximum principle and the theory of projection operator, Lions[61,62] studied the overlap domain decomposition method for the second order elliptic continuous problems. Many variations of the method were proposed to remove its sequential behavior; see, e.g., [32,43,45,63]. When it was applied to the nonselfadjoint elliptic problems discretized by the conforming finite elements, its convergence was investigated respectively in [20,24] on the basis of GMRES (generalized minimum residual) and multigrid with the conclusion that it is efficient if the subdomain diameters are small enough. Later, Chu[24] considered its applications to solving the nonselfadjoint elliptic problems discretized by the piecewise linear continuous finite elements and proved its convergence with the discrete maximum principle, not requiring that the subdomain diameters should be small enough.

This section presents the overlap domain decomposition method for the nonselfadjoint elliptic problems discretized by the Crouzeix–Raviart finite elements[27]. It is well–known that the Crouzeix–Raviart element is the simplest element, which is often adopted in scientific and engineering computing. Many practical problems, such as the convection–diffusion problem, are the nonselfadjoint elliptic problems. Their nonselfadjoint property results in some difficulties in their numerical solutions[56,57,82]. For our purpose, we first establish the discrete maximum principle, and then show the convergence of the parallel Schwarz alternating method for the nonselfadjoint elliptic problems discretized by the Crouzeix–Raviart elements. Finally, we reveal the relation between the iteration number and the overlapness of subdomains.

9.1.2 Discrete maximum principle

Consider the following nonselfadjoint elliptic problem

$$\begin{cases} -\Delta u + \beta(x) \cdot \nabla u + r(x)u = f, & \text{in } \Omega \\ u = 0, & \text{on } \partial\Omega \end{cases} \tag{9.1.1}$$

where $\Omega \subset \Re^2$ is a bounded polygonal domain, $\beta(x)$ and $r(x)$ are piecewise smooth and bounded in Ω, $r(x) \geq r_0 > 0$ and $f \in H^{-1}(\Omega)$. The variational form of (9.1.1) is

$$u \in H_0^1(\Omega) : \ a(u, v) = (f, v), \ \forall\, v \in H_0^1(\Omega) \tag{9.1.2}$$

where

$$a(u, v) = \int_\Omega \Big[\nabla u \cdot \nabla v + v\beta(x) \cdot \nabla u + r(x)uv\Big], \quad (f, v) = \int_\Omega fv.$$

By [2,35], we know that (9.1.2) has a unique solution.

Remark 9.1.1. For more general second order nonselfadjoint elliptic problem with the second order term $\sum_{i,j=1}^{2} a_{ij}(x)\frac{\partial^2 u}{\partial x_i \partial x_j}$, there always exists an inverse transformation $\begin{cases} \xi_1 = \xi_1(x_1, x_2) \\ \xi_2 = \xi_2(x_1, x_2) \end{cases}$, such that $\sum_{i,j=1}^{2} a_{ij}(x)\frac{\partial^2 u}{\partial x_i \partial x_j}$ can be transformed into $\frac{\partial^2 u}{\partial \xi_1^2} + \frac{\partial^2 u}{\partial \xi_2^2}$. Hence it is sufficient to study (9.1.1).

Suppose that $\Omega_h = \{e\}$ is a quasi–uniform mesh of Ω. Here e represents a typical triangular element. Let V_h be the Crouzeix–Raviart nonconforming finite element space[27] associated with Ω_h:

$$V_h = \Big\{v : \ v \text{ is continuous at each edge midpoint of } e, \ v|_e \text{ is linear }, \forall\, e \in \Omega_h\Big\}.$$

The Crouzeix–Raviart element discrete problem of (9.1.2) is

$$u_h \in V_h^0 : \ A(u_h, v) = (f, v), \ \forall\, v \in V_h^0 \tag{9.1.3}$$

where

$$V_h^0 = \Big\{v \in V_h : \ v(x) = 0, \ \ \forall \text{ edge midpoint } x \in \partial\Omega\Big\},$$

$$A(w, v) = \sum_{e \in \Omega_h} \int_e \Big[\nabla w \cdot \nabla v + v\beta(x) \cdot \nabla w + r(x)wv\Big].$$

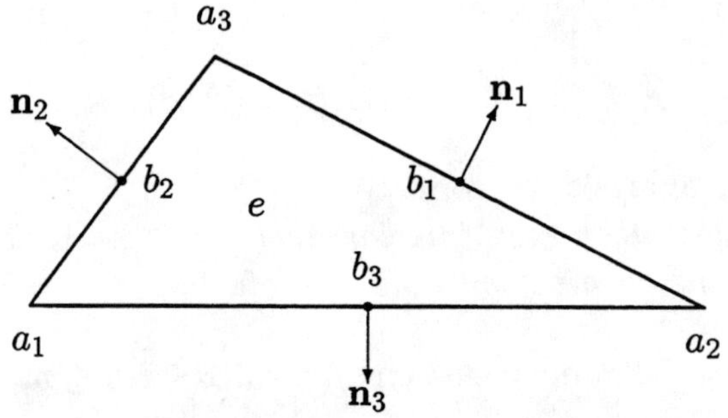

Fig.9.1.1

Let e be a triangular element as shown in Fig.9.1.1. $a_i, b_i, \mathbf{n}_i$ $(i = 1, 2, 3)$ denote the vertex, the edge midpoint and the unit outward normal vector of the opposite edge respectively. Let $|e|$ be the area of e and $\zeta_1 = |a_2a_3|, \zeta_2 = |a_3a_1|$ and $\zeta_3 = |a_1a_2|$ the lengths of its edges. λ_i $(i = 1, 2, 3)$ is the area coordinate corresponding to the vertex a_i which satisfies

$$\nabla\lambda_i = -\frac{\zeta_i}{2|e|}\mathbf{n}_i \tag{9.1.4}$$

The basis of Crouzeix–Raviart element associated with b_i is $\psi_i = 1-2\lambda_i$. Obviously, we have

Lemma 9.1.1. *Let α_0 be some positive constant. If each interior angle of $e \in \Omega_h$ is not greater than $\frac{\pi}{2} - \alpha_0$, then there exists a positive constant η, such that*

$$\int_e \nabla\psi_i \cdot \nabla\psi_j = \frac{\zeta_i\zeta_j}{|e|}\mathbf{n}_i \cdot \mathbf{n}_j \leq -\eta, \quad 1 \leq i \neq j \leq 3. \tag{9.1.5}$$

Assume that for any element $e \in \Omega_h$, each interior angle of e is not greater than $\frac{\pi}{2} - \alpha_0$, where α_0 is some positive constant. Suppose that the edge midpoints of the elements $e \in \Omega_h$ are numbered in the following order: first those in Ω, denoted by $x_1, x_2, \cdots, x_m$, then those on $\partial\Omega$, denoted by $x_{m+1}, x_{m+2}, \cdots, x_{m+l}$. Let ϕ_j $(1 \leq j \leq m+l)$ be the basis with respect to x_j. It is easy to see that

$$V_h = \mathrm{Span}\{\phi_j\}_1^{m+l}, \quad V_h^0 = \mathrm{Span}\{\phi_j\}_1^m, \quad \sum_{i=1}^{m+l}\phi_i = 1.$$

Simple calculations imply that

$$A(\phi_i, \phi_i) = O(1) + O(h) + O(h^2),$$

$$\sum_{j=1}^{m+l} A(\phi_j, \phi_i) = \sum_{e\in\Omega_h}\int_e r(x)\phi_i = O(h^2) > 0, \quad 1 \leq i \leq m.$$

If x_i and x_j are not both on the boundary of any element $e \in \Omega_h$, then $A(\phi_j, \phi_i) = 0$; if x_i and x_j are different points on the boundary of some element, then it follows from Lemma 9.1.1 that

$$A(\phi_j, \phi_i) \leq -\eta + O(h) + O(h^2).$$

Based on the above analysis, we come to

Lemma 9.1.2. *Suppose that each interior angle of $e \in \Omega_h$ is not greater than $\frac{\pi}{2} - \alpha_0$, where α_0 is some positive constant. Denote*

$$A = (a_{ij}) \in \Re^{m\times m}, \quad a_{ij} = A(\phi_j, \phi_i), \qquad 1 \leq i, j \leq m,$$
$$B = (b_{ik}) \in \Re^{m\times l}, \quad b_{ik} = A(\phi_{k+m}, \phi_i), \quad 1 \leq i \leq m, \ 1 \leq k \leq l.$$

If h is small enough, then

$$a_{ii} = O(1) > 0, \quad 1 \leq i \leq m, \tag{9.1.6}$$

$$a_{ij} \leq 0, \quad 1 \leq i \neq j \leq m, \tag{9.1.7}$$

$$\sum_{j=1}^{m} a_{ij} + \sum_{j=1}^{l} b_{ij} = \int_\Omega r(x)\phi_i = O(h^2) > 0, \quad 1 \leq i \leq m, \tag{9.1.8}$$

$$b_{ij} \leq 0, \quad 1 \leq i \leq m, \ 1 \leq j \leq l. \tag{9.1.9}$$

In what follows, $\max_{\vartheta} v$ means maximizing $v(x)$ over all the edge midpoints $x \in \vartheta$ for a measurable set ϑ.

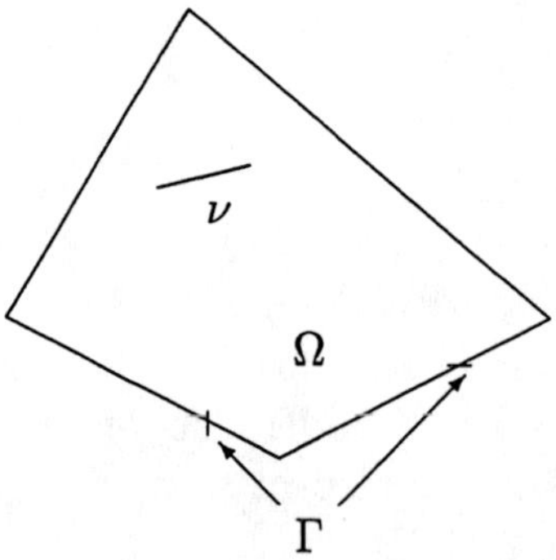

Fig.9.1.2

Lemma 9.1.3. *Suppose that $\Gamma \subset \partial\Omega, \gamma \subset \Omega, \delta = dist(\gamma, \Gamma) > 0$; see Fig.9.1.2. There exists at least one edge midpoint on Γ and γ respectively. Let $w \in V_h$ which satisfies*

$$\begin{cases} A(w, v) = 0, & \forall\, v \in V_h^0 \\ w(x) = 0, & \forall \text{ edge midpoint } x \in \partial\Omega\backslash\Gamma \\ w(x) \leq 1, & \forall \text{ edge midpoint } x \in \Gamma \end{cases} \tag{9.1.10}$$

If each interior angle of $e \in \Omega_h$ is not greater than $\frac{\pi}{2} - \alpha_0$ (α_0 is some positive constant), then there exists a constant $\mu = O(1) > 0$, independent of h, such that for sufficient small h, we have

$$\max_{\gamma} w \leq \sigma = \exp(-\mu h\delta) < 1.$$

Proof. Let $x_0 \in \gamma$ be an edge midpoint. Construct $u \in V_h$, such that

$$u(x_i) = \exp[\tilde{\mu}(|x_i - x_0| - \rho)], \quad 1 \leq i \leq m + l. \tag{9.1.11}$$

Here $\rho = \text{dist}(x_0, \Gamma)$ and $\tilde{\mu}$ is some positive constant to be determined later. Denote

$$\begin{aligned} u_i &= u(x_i), \quad i = 1, 2, \cdots, m, \qquad U = (u_1, u_2, \cdots, u_m)^T, \\ v_j &= u(x_{j+m}), \quad j = 1, 2, \cdots, l, \quad V = (v_1, v_2, \cdots, v_l)^T. \end{aligned}$$

If x_k and x_j are both on the boundary of some element, then $|x_k - x_0| \leq |x_j - x_0| + h$; otherwise,

$$a_{kj} = 0 \text{ for } 1 \leq k \neq j \leq m$$

or

$$b_{kI} = 0 \text{ for } 1 \leq k \leq m, m + 1 \leq j = I + m \leq m + l.$$

Lemma 9.1.2 gives that for $i = 1, 2, \cdots, m$

$$\begin{aligned} \sum_{j=1}^{m} a_{ij} u_j + \sum_{j=1}^{l} b_{ij} v_j &\geq u_i \Big[a_{ii} + \Big(\sum_{\substack{j=1 \\ j \neq i}}^{m} a_{ij} + \sum_{j=1}^{l} b_{ij}\Big) \exp(\tilde{\mu} h)\Big] \\ &= a_{ii} u_i \Big[1 - \Big(1 - \frac{1}{a_{ii}} \int_{\Omega} r(x) \phi_i\Big) \exp(\tilde{\mu} h)\Big]. \end{aligned}$$

In view of (9.1.6) and (9.1.8), we know that there exists a positive constant $\mu = O(1)$ such that for sufficient small h, $\tilde{\mu} = \mu h$ satisfies

$$1 - \Big(1 - \frac{1}{a_{ii}} \int_{\Omega} r(x) \phi_i\Big) \exp(\tilde{\mu} h) \geq 0.$$

Hence,

$$\sum_{j=1}^{m} a_{ij} u_j + \sum_{j=1}^{l} b_{ij} v_j \geq 0, \quad i = 1, 2, \cdots, m.$$

$$AU + BV \geq 0. \tag{9.1.12}$$

Let $w \in V_h$ be the solution of (9.1.10). Denote $W = (w(x_1), w(x_2), \cdots, w(x_m))^T$, $Z = (w(x_{m+1}), w(x_{m+2}), \cdots, w(x_{m+l}))^T$. It follows from (9.1.10) and (9.1.11) that

$$AW + BZ = 0, \tag{9.1.13}$$

$$0 \leq Z \leq V. \tag{9.1.14}$$

Lemma 9.1.2 indicates that $-B, A$ and A^{-1} are a nonnegative matrix, a M-matrix and a nonnegative matrix for sufficient small h respectively. Furthermore, using (9.1.12), (9.1.13) and (9.1.14), we have $W \leq U$. Therefore,

$$w(x_0) \leq u(x_0) = \exp(-\mu h\rho) = \exp(-\mu h \mathrm{dist}(x_0, \Gamma)).$$

By the arbitrariness of $x_0 \in \gamma$, we end the proof of Lemma 9.1.3. □

Theorem 9.1.4. (Discrete maximum principle) *Suppose that* $\Gamma \subset \partial\Omega, \gamma \subset \Omega, \delta = dist(\gamma, \Gamma) > 0$; *see Fig 9.1.2. There exists at least one edge midpoint on* Γ *and* γ *respectively. Each interior angle of e* $\in \Omega_h$ *is not greater than* $\frac{\pi}{2} - \alpha_0$ *(*α_0 *is some positive constant). Suppose that* $\tilde{w}$ *is a function defined on the edge midpoints on* Γ. *If* $w \in V_h$ *satisfies*

$$\begin{cases} A(w, v) = 0, & \forall\ v \in V_h^0 \\ w(x) = 0, & \forall\ \textit{edge midpoint}\ x \in \partial\Omega \backslash \Gamma \\ w(x) = \tilde{w}(x), & \forall\ \textit{edge midpoint}\ x \in \Gamma \end{cases}$$

then there exists a positive constant $\mu = O(1)$, *independent of* h, *such that for sufficient small* h,

$$\max_{\gamma} |w| \leq \sigma \max_{\Gamma} |\tilde{w}| = \exp(-\mu h\delta) \max_{\Gamma} |\tilde{w}|.$$

Theorem 9.1.4 is a corollary of Lemma 9.1.3.

9.1.3 Domain decomposition algorithm

Decompose Ω into $\{\Omega_i\}_1^N$, the set of the open polygonal subdomains Ω_i, such that

A1. $\Omega = \bigcup_{i=1}^{N} \Omega_i$ and $\partial\Omega_i \backslash \partial\Omega \subset \bigcup_{\substack{j=1 \\ j\neq i}}^{N} \Omega_j$, $(i = 1, 2, \cdots, N)$;

A2. If $i \neq j, j \neq k, k \neq i$, then $\partial\Omega_i \bigcap \partial\Omega_j \bigcap \partial\Omega_k = \emptyset$ and $\partial\Omega_i$ intersects $\partial\Omega_j$ at most at two points;

A3. $\partial\Omega_i$ $(i = 1, 2, \cdots, N)$ is on the gridlines of Ω_h;

A4. For any edge midpoint $x \in \Omega_h$, x is the inner point of at most $N - 1$ subdomains.

For $i = 1, 2, \cdots, N$, the subspaces of V_h associated with Ω_i are denoted as

$$V_h^i = \left\{v \in V_h : \ v(x) = 0, \ \forall \text{ edge midpoint } x \in \Omega \backslash \overline{\Omega}_i \right\},$$

$$V_h^{i,0} = \left\{v \in V_h : \ v(x) = 0, \ \forall \text{ edge midpoint } x \in \Omega \backslash \Omega_i \right\}.$$

The overlap domain decomposition method for solving (9.1.3) is described as follows.

Algorithm 9.1.1. (Parallel Schwarz alternating method)

Step 1. Choose arbitrarily $u^0 \in V_h^0$ and set $n := 0$;

Step 2. For $i = 1, 2, \cdots, N$, solve in parallel

$$\begin{cases} A(u_i^{n+1}, v) = (f, v), & \forall\, v \in V_h^{i,0} \\ u_i^{n+1} - u^n \in V_h^{i,0} \end{cases}$$

Step 3. Calculate $u^{n+1} = \frac{1}{N}\sum_{i=1}^{N} u_i^{n+1}$, set $n := n+1$ and return to step 2 until some reasonable stopping criterion is satisfied.

9.1.4 Convergence analysis

Let u_h be the solution of (9.1.3). u^n, u_i^n $(i = 1, 2, \cdots, N)$ are the iterative values of Algorithm 9.1.1 at the nth iteration. Then

$$\varepsilon^n = u^n - u_h,\ \varepsilon_i^n = u_i^n - u_h\ (i = 1, 2, \cdots, N)$$

satisfy

$$\varepsilon^n = \frac{1}{N}\sum_{i=1}^{N} \varepsilon_i^n \in V_h^0, \tag{9.1.15}$$

$$\begin{cases} A(\varepsilon_i^n, v) = 0, & \forall\, v \in V_h^{i,0} \\ \varepsilon_i^n - \varepsilon^{n-1} \in V_h^{i,0} \end{cases} \tag{9.1.16}$$

Applying (9.1.15), (9.1.16) and Theorem 9.1.4 leads to

$$\max_{\Omega_i} |\varepsilon_i^{n+1}| \le \max_{\partial\Omega_i} |\varepsilon_i^{n+1}| = \max_{\partial\Omega_i} |\varepsilon^n|,$$

$$\max_{\Omega} |\varepsilon_i^{n+1}| \le \max\Big\{\max_{\Omega\backslash\Omega_i} |\varepsilon_i^{n+1}|, \max_{\Omega_i} |\varepsilon_i^{n+1}|\Big\} \le \max_{\Omega} |\varepsilon^n|,$$

$$\max_{\Omega} |\varepsilon^{n+1}| \le \max_{\Omega} |\varepsilon^n|. \tag{9.1.17}$$

For any subdomain Ω_i, it follows from A1 and A2 that $\partial\Omega_i\backslash\partial\Omega$ can be decomposed into s $(s \le N-1)$ segments: $\Gamma_i^1, \Gamma_i^2, \cdots, \Gamma_i^s$, such that $\partial\Omega_i\backslash\partial\Omega = \bigcup_{j=1}^{s} \Gamma_i^j$ and there exists $l_j \in \{1, 2, \cdots, i-1, i+1, \cdots, N\}$, such that

$$\Gamma_i^j \subset \Omega_{l_j},\ \Gamma_i^j \bigcap (\partial\Omega_{l_j}\backslash\partial\Omega) = \emptyset.$$

Theorem 9.1.4 indicates that

$$\max_{\Gamma_i^j} |\varepsilon_{l_j}^{n+1}| \le \kappa \max_{\partial\Omega_{l_j}} |\varepsilon^n|,\ j = 1, 2, \cdots, s, \tag{9.1.18}$$

where $\kappa \in [0, 1)$ is some constant. In fact, κ is dependent on i, j and must be written as

$$\kappa_{ij} = \exp(-\mu h \text{dist}(\Gamma_i^j, \partial\Omega_{l_j}\backslash\partial\Omega)).$$

But for notational convenience, we omit the subscript of κ_{ij} here and later. It should be pointed out that although the constant κ might vary at different places, κ is always in the interval [0,1).

By (9.1.15) and (9.1.18), we get

$$\begin{aligned} N\max_{\Gamma_i^j}|\varepsilon^{n+1}| &\le \max_{\Gamma_i^j}|\varepsilon_1^{n+1}| + \cdots + \max_{\Gamma_i^j}|\varepsilon_N^{n+1}| \\ &\le (N-1)\max_{\Omega}|\varepsilon^n| + \kappa\max_{\partial\Omega_{I_j}}|\varepsilon^n| \\ &\le (N-1+\kappa)\max_{\Omega}|\varepsilon^n|. \end{aligned}$$

Hence,

$$\max_{\Gamma_i^j}|\varepsilon^{n+1}| \le \frac{N-1+\kappa}{N}\max_{\Omega}|\varepsilon^n| \le \kappa\max_{\Omega}|\varepsilon^n|,$$

$$\max_{\partial\Omega_i}|\varepsilon^{n+1}| \le \max_{1\le j\le s}(\max_{\Gamma_i^j}|\varepsilon^{n+1}|) \le \kappa\max_{\Omega}|\varepsilon^n|. \tag{9.1.19}$$

Furthermore, (9.1.17) yields

$$\max_{\bigcup_{i=1}^N \partial\Omega_i}|\varepsilon^{n+1}| \le \kappa\max_{\Omega}|\varepsilon^n| \le \kappa\max_{\Omega}|\varepsilon^{n-1}|. \tag{9.1.20}$$

For any edge midpoint $x \in \Omega\backslash\bigcup_{i=1}^N \partial\Omega_i$, there exist

$$I_1, I_2, \cdots, I_p \in \{1,2,\cdots,N\},$$

such that

$$x \in \Omega_{I_j},\ j = 1,2,\cdots,p; \quad x \notin \Omega_i,\ i \in \{1,2,\cdots,N\}\backslash\{I_1,I_2,\cdots,I_p\}.$$

It follows from (9.1.15), (9.1.17) and (9.1.19) that

$$\begin{aligned} N|\varepsilon^{n+1}(x)| &\le (N-p)|\varepsilon^n(x)| + \max_{\partial\Omega_{I_1}}|\varepsilon^n| + \cdots + \max_{\partial\Omega_{I_p}}|\varepsilon^n| \\ &\le (N-p)|\varepsilon^n(x)| + p\kappa\max_{\Omega}|\varepsilon^{n-1}|. \end{aligned}$$

A4 implies that there exist positive integers s_1 and s_2 ($s_1 + s_2 = N$), such that

$$\max_{\Omega\backslash\bigcup_{i=1}^N \partial\Omega_i}|\varepsilon^{n+1}| \le \frac{s_1}{N}\max_{\Omega}|\varepsilon^n| + \frac{s_2}{N}\kappa\max_{\Omega}|\varepsilon^{n-1}|.$$

Now applying (9.1.17) gives

$$\max_{\Omega\backslash\bigcup_{i=1}^N \partial\Omega_i}|\varepsilon^{n+1}| \le \kappa\max_{\Omega}|\varepsilon^{n-1}|.$$

Therefore, it follows from (9.1.20) that

$$\max_{\Omega}|\varepsilon^{n+1}| \le \max(\max_{\Omega\setminus\bigcup_{i=1}^{N}\partial\Omega_i}|\varepsilon^{n+1}|, \max_{\bigcup_{i=1}^{N}\partial\Omega_i}|\varepsilon^{n+1}|) \le \kappa \max_{\Omega}|\varepsilon^{n-1}|.$$

Based on the above analysis, we have

Theorem 9.1.5. *Suppose that each interior angle of $e \in \Omega_h$ is not greater than $\frac{\pi}{2} - \alpha_0$, where α_0 is some positive constant. Let u_h be the solution of (9.1.3) and u^n the iterative value of Algorithm 9.1.1 at the nth iteration. If h is small enough, then there exists a positive constant $\kappa \in [0,1)$, such that*

$$\max_{\Omega}|u^{2n} - u_h| \le \kappa^n \max_{\Omega}|u^0 - u_h|,$$

$$\max_{\Omega}|u^{2n+1} - u_h| \le \kappa^n \max_{\Omega}|u^1 - u_h|.$$

Furthermore, the higher the overlap degree, the smaller the κ.

Now we investigate Algorithm 9.1.1 from the algebraic viewpoint. Denote

$$\hat{\Omega} = \{1, 2, \cdots, m\},$$

$$\hat{\Omega}_k = \{i \in \hat{\Omega} : \text{ edge midpoint } x_i \in \Omega_k\},$$

$$A_k = (A(\phi_j, \phi_i))_{i,j\in\hat{\Omega}_k}.$$

Let m_k be the number of the elements in $\hat{\Omega}_k$. Define the restriction operator $C_k : V_h^0 \to V_h^{k,0}|_{\Omega_k}$ as

$$\forall\, v \in V_h^0,\ C_k v \in V_h^{k,0}|_{\Omega_k},\ \big(C_k v\big)(x_j) = v(x_j),\ \forall\, j \in \hat{\Omega}_k.$$

C_k is in fact a matrix of order $m_k \times m$. Similar to [89], we obtain the iterative matrix of Algorithm 9.1.1

$$G = I - \frac{1}{N}\sum_{k=1}^{N} C_k^T A_k^{-1} C_k A. \tag{9.1.21}$$

Theorem 9.1.5 indicates that if each interior angle of $e \in \Omega_h$ is not greater than $\frac{\pi}{2} - \alpha_0$ (α_0 is some positive constant) and h is small enough, Algorithm 9.1.1 converges geometrically with the spectral radius $\rho(G) < 1$. The higher the overlap degree, the smaller the κ and $\rho(G)$. In this case, Algorithm 9.1.1 converges more quickly, but the scale of each subproblem becomes larger correspondingly. It remains a problem how to decompose the domain Ω to balance the convergence rate and the computational work at each iteration.

It is known that A is a large scale sparse asymmetrical matrix (there are at most five non–zero elements in each row or column of A). Virtually, Algorithm 9.1.1 is equivalent to making the simple iteration method converge through the construction of the iterative matrix as shown in (9.1.21) by selecting the main subblocks A_i ($i =$

$1, 2, \cdots, N$) which may have common elements among them. Therefore, Algorithm 9.1.1 is efficient for solving large scale nonselfadjoint elliptic problems.

9.2 For selfadjoint problems discretized by finite elements

9.2.1 Introduction

Consider the following elliptic problem on a bounded open polygonal domain Ω in the plane

$$u \in H_0^1(\Omega): \ a(u,v) = (f,v), \ \forall\, v \in H_0^1(\Omega) \tag{9.2.1}$$

where

$$a(u,v) = \int_\Omega \left[\sum_{i,j=1}^{2} a_{ij}(x) \frac{\partial u}{\partial x_i} \frac{\partial v}{\partial x_j} + a_0(x) uv \right], \quad (f,v) = \int_\Omega fv,$$

$f \in H^{-1}(\Omega)$, $a_0(x) \geq 0$, $a_0(x)$, $a_{ij}(x)$, $(i,j = 1,2)$ are piecewise smooth and bounded functions in Ω, (a_{ij}) is a symmetric, uniformly positive definite matrix in Ω. It follows from the famous Lax–Milgram Theorem that (9.2.1) has a unique solution $u \in H_0^1(\Omega)$.

Decompose Ω into N open subdomains $\Omega_1, \Omega_2, \cdots, \Omega_N$, such that $\overline{\Omega} = \bigcup_{i=1}^{N} \overline{\Omega_i}$ and for any i, there exists j $(j \neq i)$ such that $\Omega_i \cap \Omega_j \neq \emptyset$. The overlap domain decomposition algorithm for the solving of the continuous problem (9.2.1) can be described as below:

Algorithm 9.2.1. (Parallel Schwarz alternating algorithm)

Step 1. Choose arbitrarily $u^0 \in H_0^1(\Omega)$ and set $n := 0$;

Step 2. For $i = 1, 2, \cdots, N$, solve in parallel

$$u_i^{n+1} - u^n \in V_i: \ a(u_i^{n+1}, v) = (f,v), \ \forall\, v \in V_i$$

Step 3. Calculate $u^{n+1} = \frac{1}{N} \sum_{i=1}^{N} u_i^{n+1}$, set $n := n+1$ and return to step 2.

Here $V_i = \{v \in H_0^1(\Omega) : \ \text{supp}\ v \subset \Omega_i\}$ is the subspace of $V = H_0^1(\Omega)$. Lions[61] proved that if $V = \overline{V_1 + V_2 + \cdots + V_N}$, then Algorithm 9.2.1 converges; if $V = V_1 + V_2 + \cdots + V_N$, then Algorithm 9.2.1 converges geometrically. [61] also studied how to select the subdomains $\Omega_1, \Omega_2, \cdots, \Omega_N$ such that $V = \overline{V_1 + V_2 + \cdots + V_N}$ or $V = V_1 + V_2 + \cdots + V_N$. But Algorithm 9.2.1 is of no practical use. Dryja[32] considered using Schwarz alternating method to solve the piecewise linear continuous finite element discretization of (9.2.1) and the relationship between its convergent factor and the mesh parameter. In the area of scientific/engineering computing, people often adopt the high order conforming finite elements (p–version), and the nonconforming elements to discretize (9.2.1). Therefore, it is necessary to design and analyse the Schwarz alternating algorithm for (9.2.1) discretized by the high order conforming or nonconforming finite elements.

For simplicity of the exposition, suppose Ω is a polygonal domain. We quasi–uniformly triangulate Ω into $\Omega_H = \{\hat{\Omega}_i\}_1^N$, the coarse mesh with parameter H. $S^H(\Omega)$ is the corresponding conforming finite element space on Ω_H. Then, triangulate each $\hat{\Omega}_i$ to result in the quasi–uniform mesh $\Omega_h = \{e\}$ of Ω with the parameter h. Here e represents a triangle or a quadrilateral. Finally, for each $\hat{\Omega}_i$, construct Ω_i, such that $\hat{\Omega}_i$ is the subset of Ω_i and Ω_i is compatible with Ω_h, i.e. for any $e \in \Omega_h$, $e \cap \Omega_i$ is either empty or e. In addition, suppose that

$$\text{dist}(\partial\Omega_i \backslash \partial\Omega, \partial\hat{\Omega}_i \backslash \partial\Omega) \geq L, \ i = 1, 2, \cdots, N \tag{9.2.2}$$

where L is a positive constant. L/H reflects the degree of overlapness of subdomains $\{\Omega_i\}_1^N$.

Similarly to Sect.2.3, we can get from Ω_h another mesh $\tilde{\Omega}_h = \{\tilde{e}\}$ such that each $\tilde{e} \in \tilde{\Omega}_h$ is a triangle. Let $S^h(\Omega)$ be the corresponding piecewise linear continuous finite element space associated with $\tilde{\Omega}_h$ and V_h the finite element space associated with Ω_h. π_h denotes the interpolation operator of V_h. We must emphasize here that V_h can be conforming, or nonconforming, only if it satisfies

$$\|v - \pi_h v\|_{0,e} \leq ch|v|_{1,e}, \ \forall \ v \in H^{1+\alpha}(e) \tag{9.2.3}$$

$$|\pi_h v|_{1,e} \leq c|v|_{1,e}, \ \forall \ v \in H^{1+\alpha}(e) \tag{9.2.4}$$

and there exists a linear operator $I_h : V_h \to S^h(\Omega)$ such that

$$\|v - I_h v\|_{0,e} \leq ch|v|_{1,e}, \ \forall \ v \in V_h \tag{9.2.5}$$

$$I_h v \in S_0^h(\Omega) = S^h(\Omega) \cap H_0^1(\Omega), \ \forall \ v \in V_h^0 \tag{9.2.6}$$

$$I_h v \in V_i, \ \forall \ v \in V_h^{i,0} \tag{9.2.7}$$

where

$$V_h^0 = \{v \in V_h : \ v(x) = 0, \ \forall \text{ interpolation point } x \in \partial\Omega\},$$

$$V_h^{i,0} = \{v \in V_h^0 : \ v(x) = 0, \ \forall \text{ interpolation point } x \in \Omega \backslash \Omega_i\},$$

$\alpha = 0$, if π_h is defined by the average vales (see, e.g., [53]) and $\alpha > 0$, if π_h is defined by point values (see, e.g., [27]). It is easy to see that all the conforming elements in [26] and the nonconforming elements defined in Chapter 1 satisfy (9.2.3)–(9.2.7).

9.2.2 Domain decomposition algorithm

The finite element discretization of (9.2.1) is

$$u_h \in V_h^0 : \ A(u_h, v) = (f, v), \ \forall \ v \in V_h^0 \tag{9.2.8}$$

where

$$A(w, v) = \sum_{e \subset \Omega} \int_e \left[\sum_{i,j=1}^{2} a_{ij}(x) \frac{\partial w}{\partial x_i} \frac{\partial v}{\partial x_j} + a_0(x) wv \right].$$

In what follows, we assume that there is a unique solution of (9.2.8) which converges to the solution u of (9.2.1).

Denote

$$S_0^H(\Omega) = S^H(\Omega) \cap H_0^1(\Omega), \quad |v|_{1,\Omega,h} = \Big(\sum_{e\subset\Omega} |v|_{1,e}^2\Big)^{\frac{1}{2}},$$

$$A_0(w,v) = A(w,v), \quad |v|_{1,\Omega_k,h} = \Big(\sum_{e\subset\Omega_k} |v|_{1,e}^2\Big)^{\frac{1}{2}},$$

$$A_k(w,v) = \sum_{e\subset\Omega_k} \int_e \Big[\sum_{i,j=1}^{2} a_{ij}(x)\frac{\partial w}{\partial x_i}\frac{\partial v}{\partial x_j} + a_0(x)wv\Big].$$

Algorithm 9.2.2. (Parallel Schwarz alternating algorithm)

Step 1. Choose arbitrarily $u^0 \in V_h^0$ and set $n := 0$. $\omega_0, \omega_1, \cdots, \omega_N$ are given positive parameters which satisfy $\sum_{i=0}^{N} \omega_i = 1$.

Step 2. Solve in parallel the global coarse mesh problem and the subproblem on Ω_i for $i = 1, 2, \cdots, N$,

$$\begin{cases} u_0^{n+1} \in S_0^H(\Omega) \\ A(u_0^{n+1}, v) = (f, v), \qquad \forall\, v \in S_0^H(\Omega) \end{cases}$$

$$\begin{cases} u_i^{n+1} - u^n \in V_h^{i,0} \\ A(u_i^{n+1}, v) = (f, v), \qquad \forall\, v \in V_h^{i,0} \end{cases}$$

Step 3. Calculate $u^{n+1} = \sum_{i=0}^{N} \omega_i u_i^{n+1}$, set $n := n+1$ and return to step 2 until some reasonable stopping criterion is satisfied.

9.2.3 Convergence analysis

Define the $A(\cdot,\cdot)$–projection operator $P_0 : V_h^0 \to S_0^H(\Omega)$ as follows

$$\forall\, u \in V_h^0,\ P_0 u \in S_0^H(\Omega):\ A(P_0 u, v) = A(u, v),\ \forall\, v \in S_0^H(\Omega)$$

and the $A(\cdot,\cdot)$–projection operator $P_i : V_h^0 \to V_h^{i,0}$ as follows

$$\forall\, u \in V_h^0,\ P_i u \in V_h^{i,0}:\ A(P_i u, v) = A(u, v),\ \forall\, v \in V_h^{i,0}$$

Let $Q_H : L^2(\Omega) \to S_0^H(\Omega)$ be the L^2–projection operator which satisfies

$$\forall\, u \in L^2(\Omega),\ Q_H u \in S_0^H(\Omega):\ (Q_H u, v) = (u, v),\ \forall\, v \in S_0^H(\Omega) \tag{9.2.9}$$

Lemma 9.2.1.[87] *For Q_H defined by (9.2.9), we have*

$$H^{-1}\|u - Q_H u\|_{L^2(\Omega)} + \|u - Q_H u\|_{H^1(\Omega)} \le c\|u\|_{H^1(\Omega)},\ \forall\, u \in H_0^1(\Omega)$$

Lemma 9.2.2. *We have*

$$|I_h v|_{1,e} \le c|v|_{1,e}, \ \forall\ v \in V_h \tag{9.2.10}$$

$$\|v\|_{L^2(\Omega)} \le c|v|_{1,\Omega,h}, \ \forall\ v \in V_h^0 \tag{9.2.11}$$

$$\|v\|_{L^2(\Omega_i)} \le c\sqrt{\text{meas}(\Omega_i)}|v|_{1,\Omega_i,h}, \ \forall\ v \in V_h^{i,0} \tag{9.2.12}$$

Proof. It follows from (9.2.5) and the inverse inequality implied by the quasi–uniformness of Ω_h that

$$|v - I_h v|_{1,e} \le ch^{-1}\|v - I_h v\|_{0,e} \le c|v|_{1,e},$$

which combining with the triangle inequality leads to (9.2.10).

Since $I_h v \in H_0^1(\Omega)$, we know from Lemma 1.2.1 that

$$\|I_h v\|_{L^2(\Omega)} \le c|I_h v|_{1,\Omega,h} \le c\{|v|_{1,\Omega,h} + |v - I_h v|_{1,\Omega,h}\} \le c|v|_{1,\Omega,h}.$$

Furthermore, by (9.2.5), we have

$$\|v\|_{L^2(\Omega)} \le \|v - I_h v\|_{L^2(\Omega)} + \|I_h v\|_{L^2(\Omega)} \le c|v|_{1,\Omega,h}$$

Hence, (9.2.11) holds. (9.2.12) can be shown in the same way. □

Let u_h be the solution of (9.2.8) and u^n the iterative approximate solution of Algorithm 9.2.2 at the nth iteration. Then

$$u_h - u_i^n = (I - P_i)(u_h - u^{n-1}), \ i = 0, 1, \cdots, N$$

$$u_h - u^n = (I - \sum_{i=0}^{N} \omega_i P_i)(u_h - u^{n-1}) \tag{9.2.13}$$

which is the simple iteration of the following equation

$$(\sum_{i=0}^{N} \omega_i P_i)u_h = \sum_{i=0}^{N} \omega_i f_i. \tag{9.2.14}$$

Here, $f_0, f_1, \cdots, f_N$ satisfy respectively

$$f_0 \in S_0^H(\Omega) : \ A(f_0, v) = (f, v), \ \forall\ v \in S_0^H(\Omega)$$

$$f_i \in V_h^{i,0} : \ A(f_i, v) = (f, v), \ \forall\ v \in V_h^{i,0}, \ i = 1, 2, \cdots, N$$

Denote $P = \sum_{i=0}^{N} P_i$. Then

Lemma 9.2.3.

$$A(Pv, v) \le cA(v, v), \ \forall\ v \in V_h^0$$

Proof.
$$A(P_0v, v) = A(P_0v, P_0v) \leq (A(P_0v, P_0v))^{\frac{1}{2}}(A(v,v))^{\frac{1}{2}}$$
implies that
$$A(P_0v, v) \leq A(v, v).$$
It follows from
$$A_i(P_iv, P_iv) = A(P_iv, v) = A_i(P_iv, v) \leq (A_i(P_iv, P_iv))^{\frac{1}{2}}((A_i(v,v))^{\frac{1}{2}}$$
that
$$A(P_iv, v) \leq A_i(v, v).$$
By the quasi–uniformness of the coarse mesh $\Omega_H = \{\hat{\Omega}_i\}$, we know that there is a uniform upper bound of the number of subdomains which contain x, for any $x \in \Omega$. Hence, we have
$$A(Pv, v) = A(P_0v, v) + \sum_{i=0}^{N} A(P_iv, v) \leq cA(v, v),$$
which completes the proof. □

Lemma 9.2.4.[33] *there exist $\theta_1, \theta_2, \cdots, \theta_N \in W^{1,\infty}(\Omega)$, such that*
$$\theta_i = 0 \text{ on } \Omega\backslash\Omega_i,\ \sum_{i=1}^{N} \theta_i = 1,\ 0 \leq \theta_i \leq 1,\ |\nabla\theta_i| \leq cL^{-1}, \tag{9.2.15}$$
where L is the constant in (9.2.2).

Lemma 9.2.5.[61] *Suppose that for any $v \in V_h^0$, there exist $v_0 \in S_0^H(\Omega), v_i \in V_h^{i,0}$ $(i = 1, 2, \cdots, N)$ and positive constant σ such that*
$$\sum_{i=0}^{N} v_i = v, \quad \sum_{i=0}^{N} A(v_i, v_i) \leq \sigma A(v, v).$$
Then
$$A(v, v) \leq \sigma A(Pv, v),\ \forall\ v \in V_h^0$$

Lemma 9.2.6. *We have*
$$\|v - Q_H I_h v\|_{L^2(\Omega)} \leq cH|v|_{1,\Omega,h},\ \forall\ v \in V_h^0 \tag{9.2.16}$$
$$|v - Q_H I_h v|_{1,\Omega,h} \leq c|v|_{1,\Omega,h},\ \forall\ v \in V_h^0 \tag{9.2.17}$$

Proof. For any $v \in V_h^0$, it follows from (9.2.5), Lemma 9.2.1, Lemma 1.2.1 and (9.2.10) that
$$\begin{aligned} \|v - Q_H I_h v\|_{L^2(\Omega)} &\leq \|v - I_h v\|_{L^2(\Omega)} + \|I_h v - Q_H I_h v\|_{L^2(\Omega)} \\ &\leq ch|v|_{1,\Omega,h} + CH\|I_h v\|_{H^1(\Omega)} \\ &\leq cH|v|_{1,\Omega,h} \end{aligned}$$

So, (9.2.16) is true. Similarly, (9.2.17) can be proven. □

Lemma 9.2.7. *For any $v \in V_h^0$, there exist $v_0 \in S_0^H(\Omega)$, $v_i \in V_h^{i,0}$ $(i = 1, 2, \cdots, N)$ such that*

$$\sum_{i=0}^{N} v_i = v, \quad \sum_{i=0}^{N} A(v_i, v_i) \le c(1 + L^{-2}H^2)A(v, v).$$

Proof. For any $v \in V_h^0$, let $v_0 = Q_H I_h v$, $v_i = \pi_h((v - v_0)\theta_i)$, $(i = 1, 2, \cdots, N)$, then

$$v_0 \in S_0^H(\Omega), \; v_i \in V_h^0, \; v = \sum_{i=0}^{N} v_i,$$

where $\theta_1, \theta_2, \cdots, \theta_N$ are the unit decomposition of Ω as indicated in Lemma 9.2.4.

For $i = 1, 2, \cdots, N$, it follows from Lemma 9.2.2, (9.2.4) and Lemma 9.2.4 that

$$\begin{aligned} A(v_i, v_i) &\le c|v_i|_{1,\Omega_i,h}^2 = c|\pi_h((v - v_0)\theta_i)|_{1,\Omega_i,h}^2 \\ &\le c|(v - v_0)\theta_i|_{1,\Omega_i,h}^2 \\ &\le c\{\|(v - v_0)\nabla\theta_i\|_{L^2(\Omega_i)}^2 + \|\theta_i\nabla(v - v_0)\|_{L^2(\Omega_i)}^2\} \\ &\le c\{L^{-2}\|v - v_0\|_{L^2(\Omega_i)}^2 + |v - v_0|_{1,\Omega_i,h}^2\}. \end{aligned}$$

Because there is a uniform upper bound of the number of subdomains which contain x, for any $x \in \Omega$, by Lemma 9.2.6, we have

$$\sum_{i=1}^{N} A_i(v_i, v_i) \le c\{L^{-2}\|v - v_0\|_{L^2(\Omega)}^2 + |v - v_0|_{1,\Omega,h}^2\} \le c(1 + L^{-2}H^2)|v|_{1,\Omega,h}^2.$$

On the other hand, it follows from $v_0 \in H_0^1(\Omega)$, Lemma 1.2.1 and (9.2.17) that

$$A(v_0, v_0) \le c|v_0|_{1,\Omega,h}^2 \le c|v|_{1,\Omega,h}^2$$

Therefore

$$\sum_{i=0}^{N} A_i(v_i, v_i) \le c(1 + L^{-2}H^2)|v|_{1,\Omega,h}^2 \le c(1 + L^{-2}H^2)A(v, v),$$

which completes the proof of Lemma 9.2.7. □

The main result of this section is expressed as follows:

Theorem 9.2.8. *Algorithm 9.2.2 is convergent geometrically. Its convergent factor is*

$$1 - c\frac{\omega_{min}}{1 + L^{-2}H^2},$$

where $\omega_{min} = \min_{0 \le i \le N} \omega_i$.

Proof. From (9.2.13), we know that the iterative operator of Algorithm 9.2.2 is $B = I - \sum_{i=0}^{N} \omega_i P_i$. It follows from Lemma 9.2.5 and Lemma 9.2.7 that

$$\begin{aligned} A(Bv, v) &= A(v, v) - A(\sum_{i=0}^{N} \omega_i P_i v, v) \\ &\leq A(v, v) - \omega_{min} A(Pv, v) \\ &\leq (1 - c\frac{\omega_{min}}{1 + L^{-2}H^2}) A(v, v) \end{aligned}$$

Denote $\omega_{max} = \max_{0 \leq i \leq N} \omega_i$. By Lemma 9.2.3, we have

$$A(Bv, v) \geq A(v, v) - \omega_{max} A(Pv, v) \geq (1 - c\, \omega_{max}) A(v, v).$$

Furthermore, by the Rayleigh representation theorem, the maximum and minimum eigenvalues of B satisfy respectively

$$\lambda_{max}(B) \leq 1 - c\frac{\omega_{min}}{1 + L^{-2}H^2} < 1, \quad \lambda_{min}(B) \geq 1 - c\, \omega_{max}.$$

So, the convergent factor of Algorithm 9.2.2 is

$$\rho(B) \leq 1 - c\frac{\omega_{min}}{1 + L^{-2}H^2} < 1,$$

which shows that the theorem is true. □

In general, we select $\omega_i = \frac{1}{N+1}$ in Algorithm 9.2.2 and from Theorem 9.2.8, its convergent factor is $O\Big(\frac{1}{(1+L^{-2}H^2)(1+N)}\Big)$, which means that Algorithm 9.2.2 converges independently of the finite element parameter h.

9.2.4 Matrix analysis

Let $m, m_0, m_1, \cdots, m_N$ denote the dimensions of the spaces $V_h^0, S_0^H(\Omega), V_h^{1,0}, \cdots, V_h^{N,0}$. Define the extension operator $E_0 : S_0^H(\Omega) \to V_h^0$ as: $\forall\, w \in S_0^H(\Omega)$, let $W \in \Re^{m_0}$ be the corresponding point value vector of w, then

$$E_0 W \in \Re^m, \ (E_0 W)(j) = w(x_j), \ j = 1, 2, \cdots, m,$$

where x_j is the jth inner interpolation point in Ω. Similarly, for $i = 1, 2, \cdots, N$, define the extension operator $E_i : V_h^{i,0} \to V_h^0$ as: $\forall\, w \in V_h^{i,0}$, let $W \in \Re^{m_i}$ be the corresponding point value vector of w, then

$$E_i W \in \Re^m, \ (E_i W)(j) = w(x_j), \ j = 1, 2, \cdots, m.$$

Let $A, A_0, A_1, \cdots, A_N$ be the stiff matrices of $A(\cdot, \cdot)$ with respect to V_h^0, $S_0^H(\Omega)$, $V_h^{1,0}$, $\cdots$, $V_h^{N,0}$. Let U be the point value vector of u_h, the solution of (9.2.8). Then (9.2.8) can be written in the algebraic form:

$$AU = F. \tag{9.2.18}$$

By a simple analysis, we know that the matrix representation of the operator P_i ($i = 0, 1, \cdots, N$) is $E_i A_i^{-1} E_i^T A$ and Algorithm 9.2.2 can be written as

$$U^{n+1} = (I - \sum_{i=0}^{N} \omega_i E_i A_i^{-1} E_i^T A) U^n + \sum_{i=0}^{N} \omega_i E_i A_i^{-1} E_i^T F, \tag{9.2.19}$$

which can be viewed as the simple iteration of (9.2.18) preconditioned by

$$Q = (\sum_{i=0}^{N} \omega_i E_i A_i^{-1} E_i^T)^{-1}. \tag{9.2.20}$$

From Algorithm 1.4.1, we see that if Q is used as a preconditioner to accelerate the conjugate gradient method (CG), it is unnecessary to write out the expression of Q and only Q^{-1} is involved in calculations. It is obvious that Q^{-1} is a symmetric, positive definite matrix. Q^{-1} times a vector is equivalent to the parallel solving of a global coarse mesh problem (the function of A_0^{-1}) and the fine mesh subproblem on Ω_i (the function of A_i^{-1}).

Theorem 9.2.9. *If Q is defined by (9.2.20), then*

$$Cond(Q^{-1}A) \le c(1 + L^{-2}H^2).$$

Proof. It is easy to see that for any $W \in \Re^m$

$$\frac{(AQ^{-1}AW, W)}{(AW, W)} = \frac{A(\sum_{i=0}^{N} \omega_i P_i w, w)}{A(w, w)},$$

where $w \in V_h^0$ is determined uniquely by the vector W. From Rayleigh representation theorem, $Cond(Q^{-1}A)$ is the ratio of the upper definite bound and the lower definite bound of $\frac{(AQ^{-1}AW,W)}{(AW,W)}$ over $W \in \Re^m$. Therefore, Theorem 9.2.9 follows from Lemma 9.2.3, Lemma 9.2.5 and Lemma 9.2.7. □

Theorem 9.2.9 indicates that the condition number of $Q^{-1}A$ is only related to $L^{-1}H$. If $L^{-1}H$ is kept constant, then Q is an optimal preconditioner. If L, reflecting the degree of overlapness of subdomains, is kept constant, we can improve the condition number through enlarging the scale of the global coarse mesh problem, i.e. making H smaller. On the other hand, if the global coarse mesh problem is definite, we can improve the condition number through making L larger.

Remark 9.2.1. If the global coarse mesh problem is not solved in Step 2 of Algorithm 9.2.2, then its induced preconditioner Q in (9.2.20) is changed to be

$$\tilde{Q} = (\sum_{i=1}^{N} \omega_i E_i A_i^{-1} E_i^T)^{-1}. \tag{9.2.21}$$

Similarly, we can prove that $cond(\tilde{Q}^{-1}A) \le c(1 + L^{-2})$. In general, we take $L = O(H)$, then $cond(\tilde{Q}^{-1}A) \le c(1 + H^{-2})$, which shows that the condition number is proportional to H^{-2}, when there is no global information interchange. This does not infringe the mechanism on multi–subdomain decomposition introduced by Bramble[12–15], etc.

Remark 9.2.2. Throughout our discussion by now, we have not make any regulations on N (the number of subdomains) and the shape of subdomains, which should be determined by the number of CPU of parallel computer, the differential operator in (9.2.1), the shape of the domain Ω and the finite elements.

9.2.5 Numerical experiment

Since many scholars have studied the Schwarz alternating method for the conforming finite element discretizations (see, e.g., [32,89]), we present, in this subsection, an example of application of Algorithm 9.2.2 to solve the Crouzeix–Raviart element discretization of the model problem:

$$\begin{cases} \Delta u = f, & \text{in } \Omega = (0,1)^2, \\ u = 0, & \text{on } \partial\Omega. \end{cases} \tag{9.2.22}$$

As shown in Fig.9.2.1, Ω is decomposed into four subdomains:

$$\Omega_1 = (0, 0.75)^2, \quad \Omega_2 = (0.25, 1) \times (0, 0.75),$$

$$\Omega_3 = (0.25, 1)^2, \quad \Omega_4 = (0, 0.75) \times (0.25, 1).$$

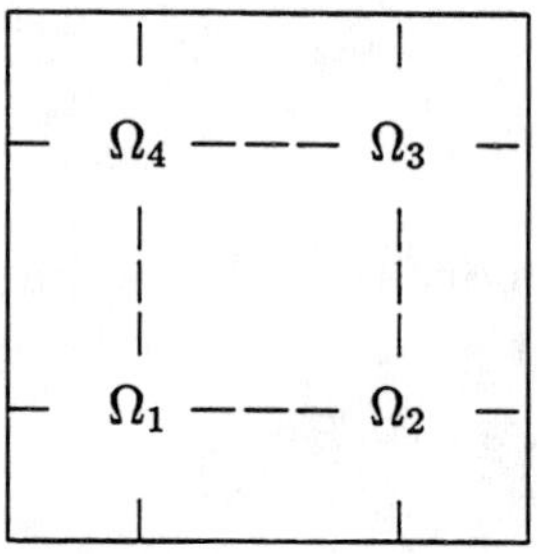

Fig.9.2.1

We triangulate Ω to get the fine mesh Ω_h such that each element $e \in \Omega_h$ is an isosceles right triangle with h as its diameter. Two methods are compared. One is CG and the other is PCG with $\tilde{Q}$ in (9.2.21) and $\omega_i = \frac{1}{4}$ as preconditioner. The iterative numbers of energy norm of error reduced by 10^{-5} are listed in Table 9.2.1, which show that the numerical result is in accordance with our theoretical analysis and there is a striking efficiency to use $\tilde{Q}$ to improve the convergence of CG method.

Table 9.2.1: A comparison between CG and PCG

h	0.2500	0.1250	0.0625
CG	12	18	31
PCG	4	5	5

Part II Domain Decomposition Methods for Elliptic Problems of Fourth Order

Chapter 10

Nonoverlap Domain Decomposition Method

10.1 Introduction

Even though the theory developed in this chapter can be applied to the plate bending problem with general boundary conditions, for simplicity we only state the results for the clamped plate. The variational problem for the clamped plate is:

$$\begin{cases} \Delta^2 u = f, & \text{in } \Omega \\ u = \dfrac{\partial u}{\partial \nu} = 0, & \text{on } \partial\Omega \end{cases} \tag{10.1.1}$$

where $\Omega \subset \Re^2$ is a polygonal domain and ν the unit outward normal vector. The variational form of (10.1.1) is

$$u \in H_0^2(\Omega): \ a(u,v) = (f,v), \ \forall\, v \in H_0^2(\Omega) \tag{10.1.2}$$

where

$$a(u,v) = \int_\Omega \Big[\Delta u \Delta v + (1-\gamma)(2\partial_{12}u\partial_{12}v - \partial_{11}u\partial_{22}v - \partial_{22}u\partial_{11}v)\Big],$$

$(f,v) = \displaystyle\int_\Omega fv$ and $\gamma \in (0, 0.5)$ is the Poisson ratio. As is well–known, the unique solvability of (10.1.2) for $f \in L^2(\Omega)$ follows from the continuity and coerciveness of the bilinear form $a(\cdot,\cdot)$ in $H_0^2(\Omega)$ (cf. [28,40,68] for details).

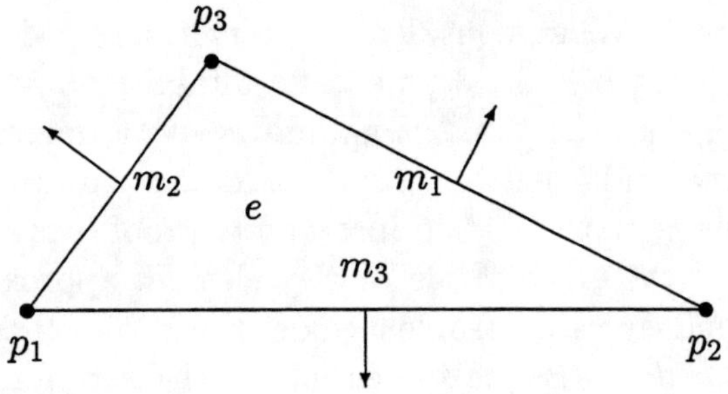

Fig.10.1.1

Suppose that $\Omega_h = \{e\}$ is a quasi–uniform mesh of Ω. Let V_h be the Morley nonconforming finite element space[67,77] associated with Ω_h. Then $v \in V_h$ if and only if it has the following three properties:

(1) $v|_e$ is quadratic, $\forall\, e \in \Omega_h$;

(2) v is continuous at each vertex p of e, $\forall\, e \in \Omega_h$;

(3) $\frac{\partial v}{\partial \nu}$ is continuous at each edge midpoint m of e, $\forall\, e \in \Omega_h$.

$v|_e$ is determined uniquely by $v(p_i)$, $\frac{\partial v}{\partial \nu}(m_i)$, $i = 1, 2, 3$; see Fig.10.1.1. Throughout this chapter, we let p and m (with or without subscripts) denote a vertex and an edge midpoint of the elements in Ω_h respectively. The Morley element discrete problem of (10.1.2) is

$$u_h \in V_h^0 : \; A(u_h, v)f = (f, v), \; \forall\, v \in V_h^0 \tag{10.1.3}$$

where

$$V_h^0 = \Big\{ v \in V_h : \; v(p) = 0, \; \frac{\partial v}{\partial \nu}(m) = 0, \; \forall\, p, m \in \partial\Omega \Big\},$$

$$A(w, v) = \sum_{e \in \Omega_h} \int_e \Big[\Delta w \Delta v + (1-\gamma)(2\partial_{12} w \partial_{12} v - \partial_{11} w \partial_{22} v - \partial_{22} w \partial_{11} v) \Big].$$

Some progress has been made in the research of domain decomposition methods for (10.1.2). Chan[22] presented interface preconditioners for the biharmonic equations via the finite difference methods where the interface consists of two grid lines, while Sun[81] constructed the multilevel preconditioners for the biharmonic equations via the B–spline methods. All their methods require that the domain Ω should be a rectangle. Brenner[19] proposed a two–level additive Schwarz preconditioner for nonconforming plate elements through intergrid transfer operators. Gu[41] considered the parallel Schwarz alternating algorithm for (10.1.3) and found the preconditioner of the algorithm by employing the idea of [32] (see the next chapter). Zhang[90−92] and Oswald[71] have recently studied hierachical, multilevel and Schwarz methods for discretizations of the biharmonic equation by conforming finite elements.

The purpose of this chapter is to give the extension theorem for Morley elements with applications to solving (10.1.3) and further to point out that the extension theorems for other plate elements[19,26] hold. It is known that the extension theorems play key roles in the analysis of nonoverlap domain decomposition methods for the second order elliptic problems discretized by the conforming or nonconforming finite element methods[41,44,46,84]. When considering the nonoverlap domain decomposition methods for the solving of (10.1.3), we must establish the extension theorem correspondingly. To this end, the conforming interpolation operator introduced in [19] is modified to act as a bridge between Morley nonconforming element space and Argyris conforming element space[4]. And its stability proof is presented thereafter. Additionally we estimate the error of the Morley element approximate solution of the inhomogeneous boundary value problem under the weak condition that the solution of (10.1.2), $u \in H^3(\Omega)$. Hence the extension theorem for Morley elements is established eventually. To illustrate its applications, we describe and analyse a nonoverlap domain decomposition algorithm with two subdomains. In each iteration of this algorithm, the solution of a discrete subproblem on one subdomain with the Dirichlet condition on the interface is followed by the solution of a discrete subproblem on another subdomain with the Neumann condition on the interface. So it is in fact the generalization of the Dirichlet–Neumann alternating method (also known as the Marini–Quarteroni algorithm[65]). Based on the extension theorem, we show that it is geometrically convergent and the convergence factor independent of h. Numerical results are also presented to indicate that the theoretical estimate is fully realized in practice. It is more important that via the same idea as above, we eventually obtain the extension theorems for all the conforming plate elements[26] and for other nonconforming plate elements[19].

The remainder of this chapter is organized as follows. In Sect.10.2, we describe and prove Theorem 10.2.4, the extension theorem for Morley elements. Its applications to the analysis of nonoverlap domain decomposition methods and numerical experiments are given in Sect.10.3. To conclude this chapter, we point out in Sect.10.4 that the extension theorems for other plate elements hold.

10.2 Extension theorem for Morley elements

The trace estimates are important tools in many nonconforming finite element analyses. For our purpose, a simple one is stated as follows:

Lemma 10.2.1.[50,60] *If e is affine equivalent to the reference element $\hat{e}$, then*

$$\int_{\partial e} w^2 \leq c\{h_e^{-1}\|w\|_{0,e}^2 + h_e|w|_{1,e}^2\},\ \forall\, w \in H^1(e)$$

Theorem 10.2.2. *Let $\tilde{\Gamma} \subset \partial\Omega$ be an open edge of a polygonal domain Ω. Suppose the functions g_1, g_2 defined on $\partial\Omega$ satisfy $g_1|_{\tilde{\Gamma}} \in H_{00}^{\frac{5}{2}}(\tilde{\Gamma})$, $g_2|_{\tilde{\Gamma}} \in H_{00}^{\frac{3}{2}}(\tilde{\Gamma})$, $g_1|_{\partial\Omega\backslash\tilde{\Gamma}} = g_2|_{\partial\Omega\backslash\tilde{\Gamma}} = 0$. Let $\theta \in H^3(\Omega), \theta_h \in V_h$ be respectively the solutions of the following problems:*

$$\begin{cases} a(\theta, v) = 0, & \forall\, v \in H_0^2(\Omega) \\ \theta = g_1, & on\ \partial\Omega \\ \frac{\partial\theta}{\partial\nu} = g_2, & on\ \partial\Omega \end{cases} \qquad \begin{cases} A(\theta_h, v) = 0, & \forall\, v \in V_h^0 \\ \theta_h(p) = g_1(p), & \forall\, p \in \partial\Omega \\ \frac{\partial\theta_h}{\partial\nu}(m) = g_2(m), & \forall\, m \in \partial\Omega \end{cases}$$

Then we have

$$|\theta - \theta_h|_{h,\Omega} \leq ch|\theta|_{H^3(\Omega)},$$

where $\qquad |w|_{h,\Omega} \triangleq \Big(\sum_{e\subset\Omega} |w|_{2,e}^2\Big)^{\frac{1}{2}}.$

Proof. Denote

$$V_h^* = \{v \in V_h :\ v(p) = \theta(p),\ \frac{\partial v}{\partial\nu}(m) = \frac{\partial\theta}{\partial\nu}(m),\ \forall\, p, m \in \partial\Omega\}.$$

For any $v \in V_h^*$, it is easy to see that

$$\begin{aligned} c|\theta_h - v|_{h,\Omega}^2 &\le A(\theta_h - v, \theta_h - v) \\ &= A(\theta - v, \theta_h - v) + A(\theta_h, \theta_h - v) - A(\theta, \theta_h - v) \\ &\le c|\theta - v|_{h,\Omega}|\theta_h - v|_{h,\Omega} + 0 + |A(\theta, \theta_h - v)|. \end{aligned}$$

Hence

$$\begin{aligned} |\theta_h - v|_{h,\Omega} &\le c\Big\{|\theta - v|_{h,\Omega} + \frac{|A(\theta,\theta_h - v)|}{|\theta_h - v|_{h,\Omega}}\Big\} \\ &\le c\Big\{|\theta - v|_{h,\Omega} + \sup_{w\in V_h^0} \frac{|A(\theta, w)|}{|w|_{h,\Omega}}\Big\}. \end{aligned}$$

By the triangle inequality, we get

$$|\theta - \theta_h|_{h,\Omega} \le c\Big(\inf_{v\in V_h^*} |\theta - v|_{h,\Omega} + \sup_{w\in V_h^0} \frac{|A(\theta, w)|}{|w|_{h,\Omega}}\Big). \tag{10.2.1}$$

(10.2.1) is in fact a variant of the second Strang lemma[26] in the nonhomogeneous boundary value case.

Let $w \in V_h^0$. Applying Green's formula yields

$$A(\theta, w) = E_1(\theta, w) + E_2(\theta, w) - \sum_{e\subset\Omega} \int_e \nabla(\Delta\theta)\cdot\nabla w, \tag{10.2.2}$$

where

$$E_1(\theta, w) = (1-\gamma)\sum_{e\subset\Omega}\int_{\partial e} \frac{\partial^2\theta}{\partial\nu\partial s}\frac{\partial w}{\partial s}, \tag{10.2.3}$$

$$E_2(\theta, w) = \sum_{e\subset\Omega}\int_{\partial e}\Big[\Delta\theta - (1-\gamma)\frac{\partial^2\theta}{\partial s^2}\Big]\frac{\partial w}{\partial \nu}. \tag{10.2.4}$$

Denote

$$\mathcal{D}(\Omega) = \{v \in \mathcal{C}^\infty(\Omega) :\ \text{supp}v \text{ is a compact subset of } \Omega\}.$$

We note that θ satisfies

$$\sum_{e\subset\Omega}\int_e \nabla(\Delta\theta)\cdot\nabla v = \int_\Omega \nabla(\Delta\theta)\cdot\nabla v = -\int_\Omega \Delta\theta\Delta v = 0,\ \forall\ v \in \mathcal{D}(\Omega)$$

Since $\mathcal{D}(\Omega)$ is dense in $H_0^1(\Omega)$, we have

$$\sum_{e\subset\Omega}\int_e \nabla(\Delta\theta)\cdot\nabla v = 0,\ \forall\ v\in H_0^1(\Omega)$$

$$A(\theta, w) = E_1(\theta, w) + E_2(\theta, w) + \sum_{e\subset\Omega}\int_e \nabla(\Delta\theta)\cdot\nabla(L_e w - w),$$

where L_e is the linear interpolation operator on e with the vertices of e as interpolation points.

Three notations: M_e, M_F and α are used throughout the remainder of this section in the following sense. M_e is the mean value operator over the element e, defined by

$$M_e v = \frac{1}{\text{meas}(e)} \int_e v, \quad \forall\, v \in L^2(e)$$

while M_F is the mean value operator over the edge F of e which can be defined similarly. $\alpha = (\alpha_1, \alpha_2) \in \mathbf{N}^2$ is a multi–index with $|\alpha| \triangleq \alpha_1 + \alpha_2$. For example, if $\alpha = (1,2)$, then $|\alpha| = 3$ and $\partial_\alpha = \dfrac{\partial^3}{\partial x \partial y^2}$.

For each edge F, if $F = \partial e \bigcap \partial\Omega$ for some e, it is obvious that $\frac{\partial}{\partial s}(L_e w) = 0$; if $F = \partial e_1 \bigcap \partial e_2$ for some elements e_1 and e_2, then $\frac{\partial}{\partial s}(L_{e_1} w)|_F = \frac{\partial}{\partial s}(L_{e_2} w)|_F$. Furthermore, for each e, $\displaystyle\int_{\partial e} \frac{\partial(w - L_e w)}{\partial s} = 0$. By the Schwarz inequality, Lemma 10.2.1, interpolation error estimates and inverse inequalities[26], we obtain

$$\begin{aligned}
E_1(\theta, w) &= (1-\gamma) \sum_{e \subset \Omega} \int_{\partial e} \Big[\frac{\partial^2 \theta}{\partial \nu \partial s} - M_e(\frac{\partial^2 \theta}{\partial \nu \partial s})\Big] \frac{\partial(w - L_e w)}{\partial s} \\
&\le (1-\gamma)\Big(\sum_{e \subset \Omega} \int_{\partial e} |\frac{\partial^2 \theta}{\partial \nu \partial s} - M_e(\frac{\partial^2 \theta}{\partial \nu \partial s})|^2\Big)^{\frac{1}{2}} \Big(\sum_{e \subset \Omega} \int_{\partial e} |\frac{\partial(w - L_e w)}{\partial s}|^2\Big)^{\frac{1}{2}} \\
&\le c\Big(\sum_{e \subset \Omega} \sum_{|\alpha|=2} \int_{\partial e} |\partial_\alpha \theta - M_e(\partial_\alpha \theta)|^2\Big)^{\frac{1}{2}} \Big(\sum_{e \subset \Omega} \sum_{|\alpha|=1} \int_{\partial e} |\partial_\alpha (w - L_e w)|^2\Big)^{\frac{1}{2}} \\
&\le c\Big(\sum_{e \subset \Omega} \sum_{|\alpha|=2} [h_e^{-1} \|\partial_\alpha \theta - M_e(\partial_\alpha \theta)\|_{0,e}^2 + h_e |\partial_\alpha \theta - M_e(\partial_\alpha \theta)|_{1,e}^2]\Big)^{\frac{1}{2}} \\
&\quad \cdot\Big(\sum_{e \subset \Omega} \sum_{|\alpha|=1} [h_e^{-1} \|\partial_\alpha (w - L_e w)\|_{0,e}^2 + h_e |\partial_\alpha (w - L_e w)|_{1,e}^2\Big)^{\frac{1}{2}} \\
&\le c\Big(\sum_{e \subset \Omega} \sum_{|\alpha|=2} h_e |\partial_\alpha \theta|_{1,e}^2\Big)^{\frac{1}{2}} \Big(\sum_{e \subset \Omega} \sum_{|\alpha|=1} h_e |\partial_\alpha w|_{1,e}^2\Big)^{\frac{1}{2}} \\
&\le ch|\theta|_{3,\Omega} |w|_{h,\Omega}.
\end{aligned}$$

We notice that for each edge F, if $F \subset \partial\Omega$, it is obvious that $M_F(\dfrac{\partial w}{\partial \nu}) = 0$; if $F = \partial e_1 \cap \partial e_2$ for some elements e_1 and e_2, then $\displaystyle\sum_{i=1}^{2} M_F(\frac{\partial w}{\partial \nu_i}|_{F \subset \partial e_i}) = 0$, where ν_i is the unit outward normal vector of e_i. In addition, $\displaystyle\int_F \Big[\frac{\partial w}{\partial \nu} - M_F(\frac{\partial w}{\partial \nu})\Big] = 0$ for each edge F and $\displaystyle\int_F (\phi - M_F \phi)^2 \le \int_F (\phi - \beta)^2$ for any measurable function ϕ and any constant $\beta \in \Re$. In the same manner as above, we have

$$\begin{aligned} E_2(\theta, w) &= \sum_{e\subset\Omega}\sum_{F\subset\partial e}\int_F \Big[\Delta\theta - (1-\gamma)\frac{\partial^2\theta}{\partial s^2} - M_F\Big(\Delta\theta - (1-\gamma)\frac{\partial^2\theta}{\partial s^2}\Big)\Big]\Big[\frac{\partial w}{\partial \nu} - M_F\Big(\frac{\partial w}{\partial \nu}\Big)\Big] \\ &\le ch|\theta|_{3,\Omega}|w|_{h,\Omega}. \end{aligned}$$

Let π_h be the interpolation operator of the Morley element space V_h. Then $\pi_h\theta \in V_h^*$. The standard interpolation error estimate gives

$$\sum_{e\subset\Omega}\int_e \nabla(\Delta\theta)\cdot\Delta(w - L_e w) \le ch|\theta|_{3,\Omega}|w|_{h,\Omega},$$

$$\inf_{v\in V_h^*}|\theta - v|_{h,\Omega} \le |\theta - \pi_h\theta|_{h,\Omega} \le ch|\theta|_{3,\Omega}.$$

By using (10.2.1), (10.2.2) and the arguments that followed, we can complete the proof of the theorem. □

Suppose that there exists an open straight line, Γ, which divides Ω into two open convex subdomains Ω_1 and Ω_2 s.t.

$$\Omega = \Omega_1 \bigcup \Omega_2 \bigcup \Gamma,\ \Omega_1 \bigcap \Omega_2 = \emptyset,\ \Gamma \bigcap e = \emptyset,\ \forall\, e \in \Omega_h.$$

For $k = 1, 2$, denote

$$|w|_{h,\Omega_k} \triangleq \Big(\sum_{e\subset\Omega_k}|w|_{2,e}^2\Big)^{\frac{1}{2}}, \qquad (w, v)_k = \int_{\Omega_k} wv,$$

$$V_h^k = \{v \in V_h^0 :\ v(p) = 0,\ \frac{\partial v}{\partial \nu}(m) = 0,\ \forall\, p, m \in \Omega\backslash\overline{\Omega}_k\},$$

$$V_h^{k,0} = \{v \in V_h^0 :\ v(p) = 0,\ \frac{\partial v}{\partial \nu}(m) = 0,\ \forall\, p, m \in \Omega\backslash\Omega_k\},$$

$$A_k(w, v) = \sum_{e\subset\Omega_k}\int_e \Big[\Delta w\Delta v + (1-\gamma)(2\partial_{12}w\partial_{12}v - \partial_{11}w\partial_{22}v - \partial_{22}w\partial_{11}v)\Big].$$

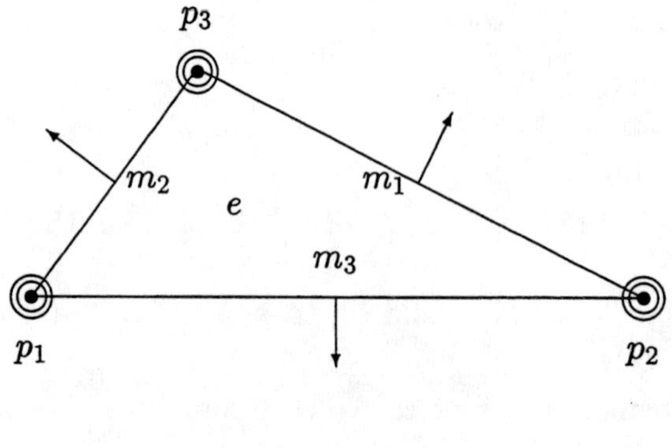

Fig.10.2.1

Brenner[19] has introduced an interpolation operator I_h^k which acts as a bridge between Morley nonconforming element space V_h^k and the Argyris conforming element[4] space $\mathcal{AR}_h^k$. Here, $v \in \mathcal{AR}_h^k$ if and only if v satisfies

(1) $v|_e$ is a fifth order polynomial, $\forall\ e \subset \Omega_k$;

(2) $\partial_\alpha v\ (0 \le |\alpha| \le 2)$ are continuous at each vertex p of e, $\forall\ e \subset \Omega_k$;

(3) $\frac{\partial v}{\partial \nu}$ is continuous at each edge midpoint m of e, $\forall\ e \subset \Omega_k$;

(4) $\partial_\alpha v(p) = 0,\ (0 \le |\alpha| \le 2),\ \frac{\partial v}{\partial \nu}(m) = 0,\ \forall\ p, m \in \partial\Omega_k \backslash \Gamma$.

$v|_e$ is determined uniquely by $\partial_\alpha v(p_i)$ and $\frac{\partial v}{\partial \nu}(m_i)$ for $0 \le |\alpha| \le 2$ and $i = 1, 2, 3$; see Fig.10.2.1. For our purpose, we modify *the conforming interpolation operator* I_h^k as follows: $\forall\ v \in V_h^k,\ I_h^k v \in \mathcal{AR}_h^k$ *s.t.*

1. $(I_h^k v)(p) = v(p)$,
2. $\dfrac{\partial(I_h^k v)}{\partial \nu}(m) = \dfrac{\partial v}{\partial \nu}(m)$,
3. $\left[\partial_\alpha (I_h^k v)\right](p) = 0, \qquad |\alpha| = 2$,
4. $\left[\partial_\alpha (I_h^k v)\right](p) = \begin{cases} 0, & |\alpha| = 1,\ p \in \partial\Omega_k \backslash \Gamma \\ \text{average of } \left(\partial_\alpha v_i\right)(p), & |\alpha| = 1,\ p \notin \partial\Omega_k \backslash \Gamma \end{cases}$

where $v_i \triangleq v|_{e_i}$ and e_i contains p as a vertex.

Theorem 10.2.3. *If I_h^k is defined as above, then*

$$I_h^k v = \frac{\partial(I_h^k v)}{\partial \nu} = 0 \quad on\ \partial\Omega_k \backslash \Gamma, \quad \forall\ v \in V_h^k \tag{10.2.5}$$

$$\|v - I_h^k v\|_{L^2(\Omega_k)} \le ch^2 |v|_{h,\Omega_k}, \ \forall\ v \in V_h^k \tag{10.2.6}$$

Proof. (10.2.5) follows from the definition of I_h^k. (10.2.6) can be obtained by modifying the proof of Lemma 5.1[19]. For completeness, we outline it as follows.

Let $v \in V_h^k$ and $e \subset \Omega_k$. Denote $w = v|_e$ and $\tilde{w} = (I_h^k v)|_e$. Then

$$w - \tilde{w} = \sum_{i=1}^{3} \sum_{|\alpha|=1,2} \partial_\alpha (w - \tilde{w})(p_i) r_{\alpha,i},$$

where the functions $r_{\alpha,i}$ are the nodal basis functions corresponding to the nodal variables $(\partial_\alpha v)(p_i)$ of the Argyris element space on e.

By the standard techniques of almost affine–equivalent family of finite elements[26], we see that

$$\|r_{\alpha,i}\|_{L^2(e)} \le ch^2 \quad \text{for } |\alpha| = 1,$$

$$\|r_{\alpha,i}\|_{L^2(e)} \le ch^3 \quad \text{for } |\alpha| = 2.$$

If $|\alpha| = 2$, then

$$|\partial_\alpha (w - \tilde{w})(p_i)| = |\partial_\alpha w(p_i)| \le |v|_{W^2_\infty(e)} \le ch^{-1} |v|_{2,e}.$$

We next discuss the case that $|\alpha| = 1$.

Suppose that $p_1 \in \partial\Omega_k\backslash\Gamma$. Since Ω_h is quasi–uniform, there exists a positive integer J, independent of h, such that $e_1, e_2, \cdots, e_J = e \subset \Omega_k$, $e_1, e_2, \cdots, e_J$ contain p_1 as a common vertex, $\text{meas}(\partial e_j \bigcap \partial e_{j+1}) > 0$ for $j = 1, 2, \cdots, J-1$ and $\text{meas}(\partial e_1 \bigcap(\partial\Omega_k\backslash\Gamma)) > 0$.

By Taylor's formula and the fact that $v|_{e_j}$ and $v|_{e_{j-1}}$ agree at the two endpoints of $\partial e_j \bigcap \partial e_{j-1}$, it is easy to obtain

$$|\frac{\partial(v|_{e_j})}{\partial s}(p_1) - \frac{\partial(v|_{e_{j-1}})}{\partial s}(p_1)| \le \frac{h}{2}\Big[|v|_{W^2_\infty(e_j)} + |v|_{W^2_\infty(e_{j-1})}\Big],$$

where s is the arc length along $\partial e_j \bigcap \partial e_{j-1}$. Similarly since $\frac{\partial(v|_{e_j})}{\partial \nu}$ and $\frac{\partial(v|_{e_{j-1}})}{\partial \nu}$ agree at the midpoint of $\partial e_j \bigcap \partial e_{j-1}$, we get

$$|\frac{\partial(v|_{e_j})}{\partial \nu}(p_1) - \frac{\partial(v|_{e_{j-1}})}{\partial \nu}(p_1)| \le \frac{h}{2}\Big[|v|_{W^2_\infty(e_j)} + |v|_{W^2_\infty(e_{j-1})}\Big].$$

Therefore, we have

$$|\partial_\alpha(v|_{e_j})(p_1) - \partial_\alpha(v|_{e_{j-1}})(p_1)| \le ch\Big[|v|_{W^2_\infty(e_j)} + |v|_{W^2_\infty(e_{j-1})}\Big].$$

Let $p_1' \in \partial e_1 \bigcap(\partial\Omega_k\backslash\Gamma)$ be another endpoint of the edge $\partial e_1 \bigcap(\partial\Omega_k\backslash\Gamma)$. Since $v(p_1) = v(p_1') = 0$, there exists a point $q \in \partial e_1 \bigcap(\partial\Omega_k\backslash\Gamma)$, s.t. $\frac{\partial(v|_{e_1})}{\partial s}(q) = 0$. Obviously, $\frac{\partial(v|_{e_1})}{\partial \nu}(m_1) = 0$, where m_1 is the midpoint of the edge $\partial e_1 \bigcap(\partial\Omega_k\backslash\Gamma)$. Then

$$|\frac{\partial(v|_{e_1})}{\partial s}(p_1)| = |\frac{\partial(v|_{e_1})}{\partial s}(p_1) - \frac{\partial(v|_{e_1})}{\partial s}(q)| \le h|v|_{W^2_\infty(e_1)},$$

$$|\frac{\partial(v|_{e_1})}{\partial \nu}(p_1)| = |\frac{\partial(v|_{e_1})}{\partial \nu}(p_1) - \frac{\partial(v|_{e_1})}{\partial \nu}(m_1)| \le h|v|_{W^2_\infty(e_1)}.$$

So

$$|\partial_\alpha(v|_{e_1})(p_1)| \le ch|v|_{W^2_\infty(e_1)}.$$

$$\begin{aligned}
|\partial_\alpha(w - \tilde{w})(p_1)| &= |\partial_\alpha(v|_e)(p_1)| \\
&= |\sum_{j=2}^{J}\Big[\partial_\alpha(v|_{e_j})(p_1) - \partial_\alpha(v|_{e_{j-1}})(p_1)\Big] + \partial_\alpha(v|_{e_1})(p_1)| \\
&\le \sum_{j=2}^{J}|\partial_\alpha(v|_{e_j})(p_1) - \partial_\alpha(v|_{e_{j-1}})(p_1)| + |\partial_\alpha(v|_{e_1})(p_1)| \\
&\le ch\sum_{j=1}^{J}|v|_{W^2_\infty(e_j)} \le c\sum_{j=1}^{J}|v|_{2,e_j} \le c\sum_{e'}|v|_{2,e'},
\end{aligned}$$

where $e' \subset \Omega_k$ s.t. $\partial e' \bigcap \partial e \ne \emptyset$.

If $p_1 \notin \partial\Omega_k\backslash\Gamma$, then by the same argument as above, we can easily obtain

$$|\partial_\alpha(w - \tilde{w})(p_1)| \le c\sum_{e'}|v|_{2,e'}.$$

Therefore, we have
$$\|v - I_h^k v\|_{L^2(e)} \le ch^2 \sum_{e'} |v|_{2,e'}.$$

Summing up the square of the last inequality over all the element $e \subset \Omega_k$, we eventually get (10.2.6) by the quasi–uniformness of the mesh Ω_h. □

In what follows, $\{p_i\}_{i=1}^I$ denotes the set of the vertices on Γ and $\{m_j\}_{j=1}^J$ the set of the edge midpoints on Γ. Let ν_k $(k = 1, 2)$ be the unit outward normal vector of Ω_k. $r_0 : V_h \to \Re^I$ and $r_1 : V_h \to \Re^J$ denote respectively the discrete operators such that
$$\forall\, v \in V_h,\ r_0 v \in \Re^I : \ \Big(r_0 v\Big)(i) = v(p_i),\ i = 1, 2, \cdots, I;$$
$$\forall\, w \in V_h,\ r_1 w \in \Re^J : \ \Big(r_1 w\Big)(j) = \frac{\partial w}{\partial \nu_1}(m_j),\ j = 1, 2, \cdots, J.$$

Define *the discrete biharmonic extension operator* $E_k^h : \Re^I \times \Re^J \to V_h^k$ as follows
$$\forall\, (\lambda, \mu) \in \Re^I \times \Re^J,\ E_k^h(\lambda, \mu) \in V_h^k : \begin{cases} A_k(E_k^h(\lambda, \mu), v) = 0, & \forall\, v \in V_h^{k,0} \\ r_0 E_k^h(\lambda, \mu) = \lambda \\ r_1 E_k^h(\lambda, \mu) = \mu \end{cases}$$

Theorem 10.2.4. (Extension theorem for Morley elements) *If Ω_1, Ω_2 are convex polygonal domains, there exist two constants σ, τ, independent of the quasi–uniform mesh parameter h, such that*
$$\sigma = \sup_{(\lambda,\mu) \in \Re^I \times \Re^J} \frac{A_1(E_1^h(\lambda, \mu), E_1^h(\lambda, \mu))}{A_2(E_2^h(\lambda, \mu), E_2^h(\lambda, \mu))} < \infty \tag{10.2.7}$$
$$\tau = \sup_{(\lambda,\mu) \in \Re^I \times \Re^J} \frac{A_2(E_2^h(\lambda, \mu), E_2^h(\lambda, \mu))}{A_1(E_1^h(\lambda, \mu), E_1^h(\lambda, \mu))} < \infty \tag{10.2.8}$$

Proof. Let $(\lambda, \mu) \in \Re^I \times \Re^J$. Denote $u_k^h = E_k^h(\lambda, \mu)$ for convenience. With the inverse inequality and Theorem 10.2.3, we have
$$|u_1^h - I_h^1 u_1^h|_{h,\Omega_1}^2 = \sum_{e \subset \Omega_1} |u_1^h - I_h^1 u_1^h|_{2,e}^2 \le ch^{-4} \|u_1^h - I_h^1 u_1^h\|_{L^2(\Omega_1)}^2 \le c|u_1^h|_{h,\Omega_1}^2.$$

Therefore, $I_h^1 u_1^h \in \mathcal{AR}_h^1 \subset H^2(\Omega_1)$ and the triangle inequality yields
$$|I_h^1 u_1^h|_{H^2(\Omega_1)} = |I_h^1 u_1^h|_{h,\Omega_1} \le c|u_1^h|_{h,\Omega_1}.$$

Furthermore, applying the trace theorem and Lemma 1.2.3, the Poincaré–Friedrichs' inequality in $H^2(\Omega_1)$ gives
$$\begin{aligned} &\|I_h^1 u_1^h\|_{H_{00}^{\frac{3}{2}}(\Gamma)}^2 + \Big\|\frac{\partial (I_h^1 u_1^h)}{\partial \nu_1}\Big\|_{H_{00}^{\frac{1}{2}}(\Gamma)}^2 \le c\Big(\|I_h^1 u_1^h\|_{H^{\frac{3}{2}}(\partial\Omega_1)}^2 + \Big\|\frac{\partial (I_h^1 u_1^h)}{\partial \nu_1}\Big\|_{H^{\frac{1}{2}}(\partial\Omega_1)}^2\Big) \\ &\le c\|I_h^1 u_1^h\|_{H^2(\Omega_1)}^2 \le c|I_h^1 u_1^h|_{H^2(\Omega_1)}^2 \le c|u_1^h|_{h,\Omega_1}^2 \le cA_1(u_1^h, u_1^h). \end{aligned} \tag{10.2.9}$$

Construct the following continuous problem:

$$\begin{cases} a(u_2, v) = 0, & \forall\, v \in H_0^2(\Omega_2) \\ u_2 = \frac{\partial}{\partial \nu_2} u_2 = 0, & \text{on } \partial\Omega_2 \backslash \Gamma \\ u_2 = I_h^1 u_1^h,\ \frac{\partial}{\partial \nu_2} u_2 = -\frac{\partial}{\partial \nu_1}(I_h^1 u_1^h), & \text{on } \Gamma \end{cases} \tag{10.2.10}$$

Note that u_2^h is the Morley approximation of u_2. By Theorem 10.2.2, we obtain

$$\begin{aligned} A_2(u_2^h, u_2^h) &\le 2\Big(A_2(u_2, u_2) + A_2(u_2 - u_2^h, u_2 - u_2^h)\Big) \\ &\le c\Big(\|u_2\|^2_{H^2(\Omega_2)} + h^2 \|u_2\|^2_{H^3(\Omega_2)}\Big) \end{aligned}$$

The well–known *a priori* inequalities of the elliptic problem (10.2.10) yield[28,40,68]

$$\begin{aligned} \|u_2\|^2_{H^2(\Omega_2)} &\le c\Big(\|u_2\|^2_{H^{\frac{3}{2}}(\partial\Omega_2)} + \|\frac{\partial}{\partial \nu_2} u_2\|^2_{H^{\frac{1}{2}}(\partial\Omega_2)}\Big), \\ \|u_2\|^2_{H^3(\Omega_2)} &\le c\Big(\|u_2\|^2_{H^{\frac{5}{2}}(\partial\Omega_2)} + \|\frac{\partial}{\partial \nu_2} u_2\|^2_{H^{\frac{3}{2}}(\partial\Omega_2)}\Big). \end{aligned}$$

Since u_2 and $\frac{\partial}{\partial \nu_2} u_2$ are piecewise polynomial on $\partial\Omega_2$, applying the fractional order inverse inequalities implied by the interpolation theorem of Sobolev spaces[1], we see that

$$|u_2|^2_{H^3(\Omega_2)} \le ch^{-2}\Big(\|u_2\|^2_{H^{\frac{3}{2}}(\partial\Omega_2)} + \|\frac{\partial}{\partial \nu_2} u_2\|^2_{H^{\frac{1}{2}}(\partial\Omega_2)}\Big).$$

With the above inequalities, we get

$$\begin{aligned} A_2(u_2^h, u_2^h) &\le c\Big(\|u_2\|^2_{H^{\frac{3}{2}}(\partial\Omega_2)} + \|\frac{\partial}{\partial \nu_2} u_2\|^2_{H^{\frac{1}{2}}(\partial\Omega_2)}\Big) \\ &\le c\Big(\|I_h^1 u_1^h\|^2_{H_{00}^{\frac{3}{2}}(\Gamma)} + \|\frac{\partial (I_h^1 u_1^h)}{\partial \nu_1}\|^2_{H_{00}^{\frac{1}{2}}(\Gamma)}\Big). \end{aligned} \tag{10.2.11}$$

Consequently, (10.2.8) follows from (10.2.9) and (10.2.11). (10.2.7) can be established in the same manner. □

Define the discrete extension operator $T_k^h : \Re^I \times \Re^J \to V_h^k$ as follows

$$\forall\, (\lambda, \mu) \in \Re^I \times \Re^J,\ T_1^h(\lambda, \mu) \in V_h^1 :\ A_1(T_1^h(\lambda, \mu), v) = -A_2(E_2^h(\lambda, \mu), v),\ \forall\, v \in V_h^0$$

$$\forall\, (\lambda, \mu) \in \Re^I \times \Re^J,\ T_2^h(\lambda, \mu) \in V_h^2 :\ A_2(T_2^h(\lambda, \mu), v) = -A_1(E_1^h(\lambda, \mu), v),\ \forall\, v \in V_h^0$$

Corollary 10.2.5. *Let $\Omega_1, \Omega_2, \sigma, \tau$ be the same as those in Theorem 10.2.4. For any $(\lambda, \mu) \in \Re^I \times \Re^J$, we have*

$$\frac{1}{\sigma} A_2(E_2^h(\lambda, \mu), E_2^h(\lambda, \mu)) \le A_1(T_1^h(\lambda, \mu), T_1^h(\lambda, \mu)) \le \tau A_2(E_2^h(\lambda, \mu), E_2^h(\lambda, \mu)), \tag{10.2.12}$$

$$\frac{1}{\tau} A_1(E_1^h(\lambda, \mu), E_1^h(\lambda, \mu)) \le A_2(T_2^h(\lambda, \mu), T_2^h(\lambda, \mu)) \le \sigma A_1(E_1^h(\lambda, \mu), E_1^h(\lambda, \mu)). \tag{10.2.13}$$

Proof. Let $(\lambda, \mu) \in \Re^I \times \Re^J$. Take $v \in V_h^0$ s.t. $v = E_k^h(\lambda, \mu)$ on Ω_k. Then by the definition of T_1^h and (10.2.7), we see that

$$\begin{aligned} A_2(E_2^h(\lambda,\mu), E_2^h(\lambda,\mu)) &= -A_1(T_1^h(\lambda,\mu), E_1^h(\lambda,\mu)) \\ &\le \left(A_1(T_1^h(\lambda,\mu), T_1^h(\lambda,\mu))\right)^{\frac{1}{2}} \left(A_1(E_1^h(\lambda,\mu), E_1^h(\lambda,\mu))\right)^{\frac{1}{2}} \\ &\le \left(A_1(T_1^h(\lambda,\mu), T_1^h(\lambda,\mu))\right)^{\frac{1}{2}} \left(\sigma A_2(E_2^h(\lambda,\mu), E_2^h(\lambda,\mu))\right)^{\frac{1}{2}}. \end{aligned}$$

So

$$A_1(T_1^h(\lambda,\mu), T_1^h(\lambda,\mu)) \ge \frac{1}{\sigma} A_2(E_2^h(\lambda,\mu), E_2^h(\lambda,\mu)).$$

Take $v \in V_h^0$ s.t.

$$v = \begin{cases} T_1^h(\lambda,\mu), & \text{on } \Omega_1 \\ E_2^h(r_0 T_1^h(\lambda,\mu), r_1 T_1^h(\lambda,\mu)), & \text{on } \Omega_2 \end{cases}$$

Then, it follows from (10.2.8) and the definition of T_1^h that

$$\begin{aligned} &A_1(T_1^h(\lambda,\mu), T_1^h(\lambda,\mu)) = -A_2(E_2^h(\lambda,\mu), E_2^h(r_0 T_1^h(\lambda,\mu), r_1 T_1^h(\lambda,\mu))) \\ &\le \left(A_2(E_2^h(\lambda,\mu), E_2^h(\lambda,\mu))\right)^{\frac{1}{2}} \\ &\quad \cdot \left(A_2(E_2^h(r_0 T_1^h(\lambda,\mu), r_1 T_1^h(\lambda,\mu)), E_2^h(r_0 T_1^h(\lambda,\mu), r_1 T_1^h(\lambda,\mu)))\right)^{\frac{1}{2}} \\ &\le \left(A_2(E_2^h(\lambda,\mu), E_2^h(\lambda,\mu))\right)^{\frac{1}{2}} \\ &\quad \cdot \left(\tau A_1(E_1^h(r_0 T_1^h(\lambda,\mu), r_1 T_1^h(\lambda,\mu)), E_1^h(r_0 T_1^h(\lambda,\mu), r_1 T_1^h(\lambda,\mu)))\right)^{\frac{1}{2}} \\ &= \left(A_2(E_2^h(\lambda,\mu), E_2^h(\lambda,\mu))\right)^{\frac{1}{2}} \left(\tau A_1(T_1^h(\lambda,\mu), T_1^h(\lambda,\mu))\right)^{\frac{1}{2}}. \end{aligned}$$

Hence we have

$$A_1(T_1^h(\lambda,\mu), T_1^h(\lambda,\mu)) \le \tau A_2(E_2^h(\lambda,\mu), E_2^h(\lambda,\mu)).$$

Combining the above inequalities yields (10.2.12).

In the same manner, (10.2.13) can be established. □

10.3 Applications to domain decomposition analysis

The extension theorem plays a key role in the analysis of nonoverlap domain decomposition algorithms with two subdomains. As an example, define an algorithm as follows:

Algorithm 10.3.1.

Step 1. Let $(\lambda^0, \mu^0) \in \Re^I \times \Re^J$ be given arbitrarily. Set n:=1.

Step 2. Find $u_1^n \in V_h^1$ by solving the subproblem on Ω_1:

$$\begin{cases} A_1(u_1^n, v) = (f, v)_1, \ \forall\, v \in V_h^{1,0} \\ r_0 u_1^n = \lambda^{n-1} \\ r_1 u_1^n = \mu^{n-1} \end{cases}$$

Step 3. Find $u_2^n \in V_h^2$ by solving the subproblem on Ω_2:

$$A_2(u_2^n, v) = -A_1(u_1^n, v) + (f, v), \quad \forall\, v \in V_h^2$$

Step 4. Select the relaxation factor $\theta_n \in (0,1)$ and calculate

$$\lambda^n = \theta_n r_0 u_2^n + (1-\theta_n)\lambda^{n-1}, \quad \mu^n = \theta_n r_1 u_2^n + (1-\theta_n)\mu^{n-1}.$$

Set n:=n+1, return to Step 2 until some reasonable stopping criterion is satisfied.

Theorem 10.3.1. *Let u_h be the solution of (10.1.3). Let $u_1^n, u_2^n, \lambda^n, \mu^n$ be the values obtained by Algorithm 10.3.1. Let $\varepsilon_k^n \in V_h^k$ s.t.*

$$\varepsilon_k^n(p) = u_k^n(p) - u_h(p), \ \frac{\partial \varepsilon_k^n}{\partial \nu}(m) = \frac{\partial u_k^n}{\partial \nu}(m) - \frac{\partial u_h}{\partial \nu}(m), \ \forall\, p, m \in \overline{\Omega}_k.$$

Denote $\delta^n = \lambda^n - r_0 u_h$ and $\eta^n = \mu^n - r_1 u_h$. Then,

(1)

$$\frac{1}{\tau} A_1(\varepsilon_1^n, \varepsilon_1^n) \le A_2(\varepsilon_2^n, \varepsilon_2^n) \le \sigma A_1(\varepsilon_1^n, \varepsilon_1^n). \tag{10.3.1}$$

(2) There exists a constant $\theta^* \in (0,1]$, such that

$$A_1(\varepsilon_1^{n+1}, \varepsilon_1^{n+1}) \le \kappa(\theta_n) A_1(\varepsilon_1^n, \varepsilon_1^n), \tag{10.3.2}$$

where $\kappa(\theta_n) < 1, \ \forall\, \theta_n \in (0, \theta^*)$.

(3) There exists the optimal relaxation factor θ^{opt}, such that

$$\kappa(\theta^{opt}) = \min_{\theta \in (0,\theta^*)} \kappa(\theta). \tag{10.3.3}$$

Proof. It is easy to see that $\varepsilon_k^{n+1} \in V_h^k$ satisfies

$$\begin{cases} A_1(\varepsilon_1^{n+1}, v) = 0, \ \forall\, v \in V_h^{1,0} \\ r_0 \varepsilon_1^{n+1} = \delta^n \\ r_1 \varepsilon_1^{n+1} = \eta^n \end{cases} \tag{10.3.4}$$

$$A_2(\varepsilon_2^{n+1}, v) = -A_1(\varepsilon_1^{n+1}, v), \ \forall\, v \in V_h^2 \tag{10.3.5}$$

$$\delta^{n+1} = \theta_{n+1} r_0 \varepsilon_2^{n+1} + (1-\theta_{n+1})\delta^n, \quad \eta^{n+1} = \theta_{n+1} r_1 \varepsilon_2^{n+1} + (1-\theta_{n+1})\eta^n. \tag{10.3.6}$$

(10.3.4) and (10.3.5) yield

$$A_2(\varepsilon_2^{n+1}, v) = -A_1(\varepsilon_1^{n+1}, v), \ \forall\, v \in V_h^0 \tag{10.3.7}$$

Therefore

$$\varepsilon_1^{n+1} = E_1^h(\delta^n, \eta^n), \ \varepsilon_2^{n+1} = T_2^h(\delta^n, \eta^n).$$

By Corollary 10.2.5, we get (10.3.1).

Furthermore, it follows from (10.3.4) and (10.3.6) that

$$\begin{aligned} \varepsilon_1^{n+1} &= \theta_n E_1^h(r_0\varepsilon_2^n, r_1\varepsilon_2^n) + (1-\theta_n)\varepsilon_1^n, \\ A_1(\varepsilon_1^{n+1}, \varepsilon_1^{n+1}) = & \ \theta_n^2 A_1(E_1^h(r_0\varepsilon_2^n, r_1\varepsilon_2^n), E_1^h(r_0\varepsilon_2^n, r_1\varepsilon_2^n)) \\ & +2\theta_n(1-\theta_n)A_1(E_1^h(r_0\varepsilon_2^n, r_1\varepsilon_2^n), \varepsilon_1^n) \\ & +(1-\theta_n)^2 A_1(\varepsilon_1^n, \varepsilon_1^n). \end{aligned} \tag{10.3.8}$$

By (10.2.7) and (10.3.1), we see that

$$A_1(E_1^h(r_0\varepsilon_2^n, r_1\varepsilon_2^n), E_1^h(r_0\varepsilon_2^n, r_1\varepsilon_2^n)) \le \sigma A_2(\varepsilon_2^n, \varepsilon_2^n) \le \sigma^2 A_1(\varepsilon_1^n, \varepsilon_1^n). \tag{10.3.9}$$

(10.3.7) gives

$$A_2(\varepsilon_2^{n+1}, \varepsilon_2^{n+1}) = -A_1(\varepsilon_1^{n+1}, E_1^h(r_0\varepsilon_2^{n+1}, r_1\varepsilon_2^{n+1})).$$

So by (10.3.1), we get

$$A_1(E_1^h(r_0\varepsilon_2^n, r_1\varepsilon_2^n), \varepsilon_1^n) = -A_2(\varepsilon_2^n, \varepsilon_2^n) \le -\frac{1}{\tau} A_1(\varepsilon_1^n, \varepsilon_1^n). \tag{10.3.10}$$

If $0 < \theta_n < 1$, then (10.3.2) follows from (10.3.8), (10.3.9) and (10.3.10). Furthermore, $\kappa(\theta_n)$ has the following expression

$$\kappa(\theta_n) = \frac{1}{\tau}\Big(\theta_n^2(\sigma^2\tau + \tau + 2) - 2\theta_n(\tau+1) + \tau\Big). \tag{10.3.11}$$

An elementary calculation indicates $0 \le \kappa(\theta_n) < 1$, if and only if

$$0 < \theta_n < \theta^* = \min\Big(1, \frac{2(\tau+1)}{\sigma^2\tau + \tau + 2}\Big).$$

The optimal relaxation factor is given by

$$\theta^{opt} = \frac{\tau+1}{\sigma^2\tau + \tau + 2}, \tag{10.3.12}$$

for which

$$\kappa(\theta^{opt}) = \frac{\sigma^2\tau^2 - 1}{\tau(\sigma^2\tau + \tau + 2)} = \min_{\theta \in (0, \theta^*)} \kappa(\theta). \tag{10.3.13}$$

So (10.3.3) follows from (10.3.13). □

Algorithm 10.3.1 is the generalization of the so-called Dirichlet–Neumann alternative method or the Marini–Quarteroni method[65] to plate bending problems. Its essence lies in the continuing correction of the initial guess of (r_0u_h, r_1u_h) by imposing $r_0(\Delta u_1^n) = r_0(\Delta u_2^n)$ and $r_1(\Delta u_1^n) = r_1(\Delta u_2^n)$ at each iteration. Theorem

10.3.1 implies that Algorithm 10.3.1 converges geometrically and independently of h, which is guaranteed by the extension theorem (Theorem 10.2.4). In the special case that the domain Ω is symmetric with respect to Γ, then $\sigma = \tau = 1$ in (10.2.7) and (10.2.8), thus, by (10.3.12) and (10.3.13), $\theta^{opt} = \frac{1}{2}$ and $\kappa(\theta^{opt}) = 0$, which together with (10.3.3) show that only one iteration is needed to obtain the solution of (10.1.3).

Of course, other algorithms in Chapter 3 can be generalized and their analysis can be carried out similarly, based on Theorem 10.2.4.

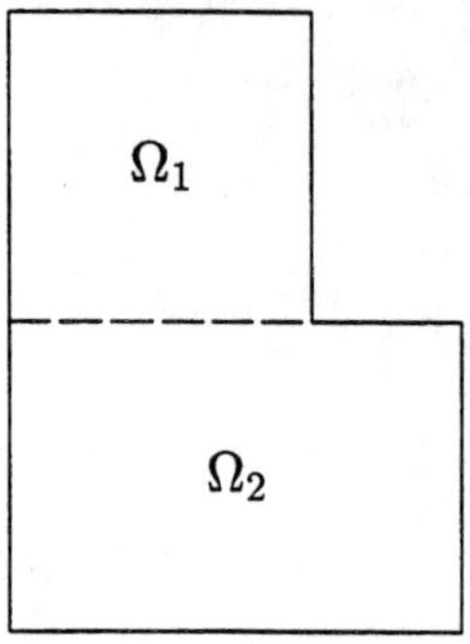

Fig.10.3.1

To end this section, we present a numerical experiment. Decompose the domain $\Omega = (0, 1.5) \times (0, 1) \bigcup (0, 1) \times [1, 2)$ into subdomains: $\Omega_1 = (0, 1) \times (1, 2)$, $\Omega_2 = (0, 1.5) \times (0, 1)$. See Fig.10.3.1. Triangulate Ω to get the fine mesh Ω_h so that each element $e \in \Omega_h$ is an isosceles right triangle with h as its diameter. When $h = 0.25$, there are 32, 48 elements and 81, 117 interpolation points in Ω_1, Ω_2, respectively. When $h = 0.125$, there are 128, 192 elements and 289, 425 interpolation points in Ω_1, Ω_2, respectively. When $h = 0.0625$, there are 512, 768 elements and 1089, 1617 interpolation points in Ω_1, Ω_2, respectively. In the above three cases, there are 7, 15 and 31 interpolation points on Γ respectively. For an edge midpoint m, if $m \in \partial e_1 \bigcap \partial e_2$, then the outward normal vectors of e_1 and e_2 at m are opposite. To ensure that $\frac{\partial v}{\partial \nu}(m)$ are determined uniquely, we require that the outward (inward) normal vectors should be chosen for the triangular elements with even (odd) numbers. In the following tables, n is the number of iterations, ε^n is the error after n iterations, $\|\varepsilon^n\|_A = A(\varepsilon^n, \varepsilon^n)$, $\rho_n = \sqrt[n]{\|\varepsilon^n\|_A / \|\varepsilon^0\|_A}$ and $\|\varepsilon^n\|_\infty = \|\varepsilon^n\|_{L^\infty(\Omega)}$.

When using Algorithm 10.3.1 to solve (10.1.3), a procedure is built up to generate a sequence of the discrete biharmonic functions on Ω_1 and Ω_2 with the same values at $p, m \in \Gamma$. This allows us to compute, at each iteration, two constants σ_n, τ_n suggested by (10.2.7) and (10.2.8), which combined with (10.3.12) gives the sequence of approximate values θ_n of the optimal relaxation factor θ^{opt}. We point out that the evaluation of θ_n does not require the solution of any additional problem in our algorithm (for details, cf.[65]). The main experimental results, obtained on a SGI work station, are listed in Table 10.3.1 and Table 10.3.2, and support our theoretical analysis.

Table 10.3.1: Error reduction factor ρ_n vs. h

n	1	5	9	13
$h = 0.2500$	0.0253	0.0343	0.0261	0.0395
$h = 0.1250$	0.0272	0.0476	0.0438	0.0558
$h = 0.0625$	0.0264	0.0513	0.0473	0.0464

Table 10.3.2: The errors $\|\varepsilon^n\|_A$ and $\|\varepsilon^n\|_\infty$ when $h = 0.0625$

n	1	5	9	13
$\|\varepsilon^n\|_A$	$0.487 \cdot 10^9$	$0.129 \cdot 10^4$	$0.532 \cdot 10^{-1}$	$0.718 \cdot 10^{-6}$
$\|\varepsilon^n\|_\infty$	$0.952 \cdot 10^4$	$0.837 \cdot 10^2$	$0.261 \cdot 10^0$	$0.459 \cdot 10^{-3}$

10.4 Extension theorems for other plate elements

Let V_h be Fraeijs de Veubeke element space $\mathcal{F}_h$ or Zienkiewicz element space $\mathcal{Z}_h$ or Adini element space $\mathcal{AD}_h$ described in [19]. As in Sect.10.2, we can define *the discrete biharmonic operator* $E_k^h : \Re^M \to V_h^k$ correspondingly. Here M denotes the number of degrees of freedom associated with the interface Γ.

Theorem 10.4.1. (Extension theorem) *Let Ω_1, Ω_2 be convex. If V_h is one of the three nonconforming plate element spaces $\mathcal{F}_h$, $\mathcal{Z}_h$ and $\mathcal{AD}_h$, there exist two constants $\hat{\sigma}, \hat{\tau}$, independent of the quasi-uniform mesh parameter h, such that*

$$\hat{\sigma} = \sup_{\eta \in \Re^M} \frac{A_1(E_1^h\eta, E_1^h\eta)}{A_2(E_2^h\eta, E_2^h\eta)} < \infty, \quad \hat{\tau} = \sup_{\eta \in \Re^M} \frac{A_2(E_2^h\eta, E_2^h\eta)}{A_1(E_1^h\eta, E_1^h\eta)} < \infty. \tag{10.4.1}$$

We can adopt the ideas of the proof of Theorem 10.2.4 to prove Theorem 10.4.1 with the following points in mind:

1. The error of the nonconforming approximate solution of the inhomogeneous boundary value problem can be estimated by first obtaining an inequality similar to (10.2.1), subtracting off appropriate 'conforming' parts as in (10.2.2) and then applying the bilinear lemma[26], cf. [5,26,59,79].

2. *The conforming interpolation operator* I_h^k must be constructed by similarly modifying the corresponding one introduced in [19] and Theorem 10.2.3 still holds in this case.

Theorem 10.4.2. (Extension theorem) *Let Ω_1, Ω_2 be convex. If V_h is one of the conforming plate element spaces[26], then (10.4.1) holds.*

Since it is unnecessary to construct the conforming interpolation operator I_h^k in this case, the proof of Theorem 10.4.2 is much simpler than the proof of Theorem 10.2.4, so we omit it here. Analogous results may be found in [84].

Chapter 11

Overlap Domain Decomposition Methods

11.1 Introduction

Consider the variational form of the Dirichlet boundary value problem:

$$u \in V: \quad a(u,v) = (f,v), \quad \forall\, v \in V \tag{11.1.1}$$

where $V = H_0^2(\Omega)$, $(f,v) = \int_\Omega fv$, $a(u,v)$ is a symmetric, continuous, $H_0^2(\Omega)$–positive definite bilinear form, e.g.

$$a(u,v) = \int_\Omega \Delta u \Delta v, \tag{11.1.2}$$

and

$$a(u,v) = \int_\Omega \Big[\Delta u \Delta v + (1-\sigma)(2\frac{\partial^2 u}{\partial x_1 \partial x_2}\frac{\partial^2 v}{\partial x_1 \partial x_2} - \frac{\partial^2 u}{\partial x_1^2}\frac{\partial^2 v}{\partial x_2^2} - \frac{\partial^2 u}{\partial x_2^2}\frac{\partial^2 v}{\partial x_1^2})\Big]. \tag{11.1.3}$$

Ω is a plane open domain. (11.1.2) arises from the Stokes problem in fluid mechanics[26], etc., while (11.1.3) comes from the plate bending problem in structure mechanics[26], etc. Here, $\sigma \in [0, 0.5]$ is the Poisson ratio. It follows from [28,40,68] that if $f \in L^2(\Omega)$ then (11.1.1) has and only has one solution. Since there is little difference to cope with (11.1.2) and (11.1.3), we study the Schwarz alternating method only with (11.1.2) in this chapter.

Suppose that $\Omega_h = \{e\}$ is a quasi–uniform mesh of Ω. Let V_h be the Morley nonconforming finite element space on Ω_h; see Sect.10.1. The Morley finite element discretization of variational problem (11.1.1) is

$$u_h \in V_h^0: \quad A(u_h,v) = (f,v), \quad \forall\, v \in V_h^0 \tag{11.1.4}$$

where

$$V_h^0 = \Big\{v \in V_h: \ v(p) = 0, \ \frac{\partial v}{\partial \nu}(m) = 0, \ \forall\, p, m \in \partial\Omega\Big\},$$

$$A(w,v) = \sum_{e\in\Omega_h} \int_e \Delta w \Delta v.$$

Detailed studies have been carried out on the Schwarz alternating method for the solving of second order elliptic problems dicretized by finite elements in [32,61,63]. The aim of this chapter is to generalize this method to the solving of (11.1.4).

11.2 Algorithm for the continuous problem

Applying Theorem 1.5[2] and introducing mollification kernel, we can show

Lemma 11.2.1. *Suppose D and G are bounded open domains in the plane such that $G \subset\subset D$. Then there exist $\theta \in C_0^\infty(D)$ s.t.*

$$\begin{cases} 0 \le \theta(x) \le 1, & \forall\, x \in \Re^2 \\ \theta(x) = 1, & \forall\, x \in \overline{G} \\ |D_{l_1}^n \theta(x)| \le c\delta^{-n}, & \forall\, x \in D \end{cases}$$

where $n \ge 1, \delta = dist(\partial G, \partial D), D_{l_1}^n \theta = \dfrac{\partial^n \theta}{\partial x_1^{l_1} \partial x_2^{l_2}}$, $l_1 \in [0, n]$ is an integer, $l_2 = n - l_1$. $G \subset\subset D$ means that G is the relative compact true subset of D.

Lemma 11.2.2. *Let the bounded open domains $D, G_1, G_2, \cdots, G_N \subset \Re^2$ satisfy $\overline{D} \subset \bigcup_{i=1}^{N} G_i$. Assume there is a uniform upper bound of the number of the domains G_i which contain x, for any $x \in D$. If there exists a relative compact true subset of G_i, $G_i^* \subset\subset G_i$ such that $\overline{D} \subset \bigcup_{i=1}^{N} \overline{G_i^*}$, then there exist $\theta_1, \theta_2, \cdots, \theta_N$, called the unit decomposition of D, such that*

$$\begin{cases} 0 \le \theta_i(x) \le 1, & \forall\, x \in \Re^2 \\ \theta_i \in C_0^\infty(G_i) & \\ \sum\limits_{i=1}^{N} \theta_i(x) = 1, & \forall\, x \in \overline{D} \\ \max\limits_{x \in \Re^2} |(D_{l_1}^n \theta_i)(x)| \le c\delta^{-n} & \end{cases}$$

where, $\delta = \min\limits_{1 \le i \le N} dist(\partial G_i, \partial G_i^*) > 0$.

Lemma 11.2.2 is a corollary of Lemma 11.2.1.

Lemma 11.2.3.[61] *Let W_i be the close subspace of the Hilbert space, such that $W = \sum_{i=1}^{N} W_i$. Then there exists a positive constant β, such that for any $w \in W$, there exist $w_1 \in W_1$, $w_2 \in W_2$, $\cdots$, $w_N \in W_N$ which satisfy*

$$w = \sum_{i=1}^{N} w_i, \quad \sum_{i=1}^{N} \|w_i\|^2 \le \beta^2 \|w\|^2,$$

where $\|\cdot\|$ represents the induced norm from the inner product $(\cdot,\cdot)$ of the Hilbert sapce W.

Define $P_i : W \to W_i$, the projection operator from the Hilbert sapce W to its close subspace W_i as

$$\forall\, w \in W,\ P_i w \in W_i :\ (P_i w, v) = (w, v),\ \forall\, v \in W_i \qquad (11.2.1)$$

By Lemma 11.2.3 and the Schwarz inequality, we can get

Lemma 11.2.4. *Let $W, W_i, \beta, (\cdot,\cdot), \|\cdot\|$ be the same as those in Lemma 11.2.3 and P_i the projection operator defined by (11.2.1). Then*

$$\|w\|^2 \le \beta^2 \sum_{i=1}^{N} \|P_i w\|^2, \quad \forall\, w \in W$$

Decompose Ω overlaply into N open subdomains $\Omega_1, \Omega_2, \cdots, \Omega_N$ such that $\Omega = \bigcup_{k=1}^{N} \Omega_k$ and for each Ω_k, there exists $\Omega_k^* \subset \Omega_k$ s.t.

$$\mathrm{dist}(\partial\Omega_k \backslash \partial\Omega, \partial\Omega_k^* \backslash \partial\Omega) \ge d, \quad \overline{\Omega} = \bigcup_{k=1}^{N} \overline{\Omega_k^*}, \tag{11.2.2}$$

where d is a positive constant.

Denote

$$V_k = \{v \in H_0^2(\Omega) : \ \mathrm{supp}\ v \in H_0^2(\Omega_k)\}.$$

It follows from the construction of $\{\Omega_k\}_1^N$ that $V = \sum_{k=1}^{N} V_k$. Now we generalize the Schwarz alternating method in [32] to solve (11.1.1).

Algorithm 11.2.1.

Step 1. Choose arbitrarily $u^0 \in V$ and set $n := 1$;

Step 2. For $k = 1, 2, \cdots, N$, solve $u_k^n \in V$ in parallel

$$\begin{cases} u_k^n - u^{n-1} \in V_k \\ a(u_k^n, v) = (f, v), \quad \forall\, v \in V_k \end{cases}$$

Step 3. Calculate $u^n = \frac{1}{N} \sum_{k=1}^{N} u_k^n$, set $n := n + 1$ and return to step 2.

Similarly to (11.2.1), we can define the $a(\cdot,\cdot)$–projection operator from V to V_k, which might be denoted as P_k as well. Then Algorithm 11.2.1 can be written as

$$u^{n+1} - u = (I - \frac{1}{N} \sum_{k=1}^{N} P_k)(u^n - u), \tag{11.2.3}$$

Where u is the solution of (11.1.1).

Next we give the estimate of the β in Lemma 11.2.3. Of course, $a(\cdot,\cdot)$ is an inner product on $V = H_0^2(\Omega)$. Its induced norm is denoted as $\|\cdot\|$ for convenience. It follows from (11.2.2) and Lemma 11.2.2 that there exist $\theta_1, \theta_2, \cdots, \theta_N$, such that

$$\begin{cases} 0 \le \theta_k(x) \le 1, \quad \forall\, x \in \Omega \\ \theta_k(x) = 0, \quad \forall\, x \in \Omega \backslash \Omega_k \\ \sum_{k=1}^{N} \theta_k(x) = 1, \quad \forall\, x \in \Omega \\ \max_{x \in \Re^2} |(D_{l_1}^n \theta_k)(x)| \le c d^{-n}, \ n = 1, 2, \ l_1 = 0, 1, \cdots, n \end{cases}$$

where d is the same as the one in (11.2.2).

For any $v \in V$, let $v_k = \theta_k v$. Then $v_k \in V_k$ and

$$\begin{aligned}\int_{\Omega_k} (\frac{\partial^2 v_k}{\partial x_i \partial x_j})^2 &= \int_{\Omega_k} \Big[\frac{\partial v}{\partial x_i}\frac{\partial \theta_k}{\partial x_j} + \frac{\partial v}{\partial x_j}\frac{\partial \theta_k}{\partial x_i} + v\frac{\partial^2 \theta_k}{\partial x_i \partial x_j} + \theta_k \frac{\partial^2 v}{\partial x_i \partial x_j}\Big]^2 \\ &\leq c\int_{\Omega_k} \Big[d^{-2}(\frac{\partial v}{\partial x_i})^2 + d^{-2}(\frac{\partial v}{\partial x_j})^2 + d^{-4}v^2 + (\frac{\partial^2 v}{\partial x_i \partial x_j})^2\Big] \\ &\leq c[d^{-4}\|v\|^2_{H^0(\Omega_k)} + d^{-2}|v|^2_{H^1(\Omega_k)} + |v|^2_{H^2(\Omega_k)}].\end{aligned}$$

Therefore

$$\begin{aligned}\|v_k\|^2 &= \int_{\Omega} \sum_{i,j=1}^{2} (\frac{\partial^2 v_k}{\partial x_i \partial x_j})^2 = \int_{\Omega_k} \sum_{i,j=1}^{2} (\frac{\partial^2 v_k}{\partial x_i \partial x_j})^2 \\ &\leq c[d^{-4}\|v\|^2_{H^0(\Omega_k)} + d^{-2}|v|^2_{H^1(\Omega_k)} + |v|^2_{H^2(\Omega_k)}].\end{aligned}$$

Since there is a uniform upper bound of the number of the domains Ω_k which contain x, for any $x \in \Omega$, we have

$$\sum_{k=1}^{N} \|v_k\|^2 \leq c[d^{-4}\|v\|^2_{H^0(\Omega)} + d^{-2}|v|^2_{H^1(\Omega)} + |v|^2_{H^2(\Omega)}].$$

Furthermore, it follows from $v \in V = H_0^2(\Omega)$ and Lemma 1.2.1 that

$$\sum_{k=1}^{N} \|v_k\|^2 \leq c(1 + d^{-4})\|v\|^2. \tag{11.2.4}$$

So, under (11.2.2), the β in Lemma 11.2.3 is $c(1 + d^{-2})$.

Theorem 11.2.5. *(11.2.3) is convergent and there exists a positive constant $\alpha \in [0, 1)$ such that*

$$\|I - \frac{1}{N}\sum_{k=1}^{N} P_k\| \leq \alpha.$$

Proof. $\forall\, v \in V$, the definition of P_k yields

$$\|v\|^2 = \|P_k v\|^2 + \|(I - P_k)v\|^2,$$

$$N\|v\|^2 = \sum_{k=1}^{N} \|P_k v\|^2 + \sum_{k=1}^{N} \|(I - P_k)v\|^2.$$

It follows from (11.2.4), Lemma 11.2.3 and Lemma 11.2.4 that

$$\frac{1}{N}\sum_{k=1}^{N} \|(I - P_k)v\|^2 = \|v\|^2 - \frac{1}{N}\sum_{k=1}^{N} \|P_k v\|^2 \leq (1 - \frac{1}{cN(1 + d^{-4})})\|v\|^2,$$

Furthermore, by the Schwarz inequality, we get

$$\|(I - \frac{1}{N}\sum_{k=1}^{N} P_k)v\| \leq \Big[\frac{1}{N}\sum_{k=1}^{N} \|(I - P_k)v\|\Big]^2 \leq (1 - \frac{1}{cN(1 + d^{-4})})\|v\|^2,$$

Taking $\alpha = \sqrt{1 - \frac{1}{cN(1+d^{-4})}}$ completes the proof. □

Theorem 11.2.5 indicates that Algorithm 11.2.1 converges geometrically and the convergent rate depends on the degree of overlapness of subdomains. The larger the d in (11.2.2), the higher the degree of overlapness of subdomains, the faster it converges.

11.3 Algorithm for the discrete problem

Let $\Omega_H = \{\hat{\Omega}_k\}_1^N$ be the quasi–uniform coarse mesh of Ω with the parameter H ($H \gg h$), which is compatible with the fine mesh $\Omega_h = \{e\}$, i.e.

$$\hat{\Omega}_k \cap e = \emptyset, \ \forall\, \hat{\Omega}_k \in \Omega_H, \ \forall\, e \in \Omega_h.$$

Then for each $\hat{\Omega}_k$, construct Ω_k s.t. $\partial\Omega_k$ is on the gridlines of Ω_h, $\hat{\Omega}_k \subset \Omega_k$ and

$$\text{dist}(\partial\Omega_k \backslash \partial\Omega, \partial\hat{\Omega}_k \backslash \partial\Omega) \geq d, \quad k = 1, 2, \cdots, N \tag{11.3.1}$$

Denote

$$V_h^{k,0} = \{v \in V_h^0 : \ v(p) = 0, \frac{\partial v}{\partial \nu}(m) = 0, \forall\, p, m \in \Omega \backslash \Omega_k\}.$$

The parallel Schwarz alternating method for the solving of (11.1.4) is described as

Algorithm 11.3.1.

Step 1. Choose arbitrarily $u^0 \in V_h^0$ and set $n := 1$;

Step 2. For $k = 1, 2, \cdots, N$, solve $u_k^n \in V_h^0$ in parallel

$$\begin{cases} u_k^n - u^{n-1} \in V_h^{k,0} \\ A(u_k^n, v) = (f, v), \quad \forall\, v \in V_h^{k,0} \end{cases}$$

Step 3. Calculate $u^n = \dfrac{1}{N} \displaystyle\sum_{k=1}^N u_k^n$, set $n := n + 1$ and return to step 2.

Let m and m_k represent the dimensions of V_h^0 and $V_h^{k,0}$. x_j denotes the jth inner interpolation point in Ω. Define the extension operator $E_k : V_h^{k,0} \to V_h^0$ as

$$\forall\, w_k \in V_h^{k,0}, \ E_k w_k \in V_h^0, \text{ the freedoms of } w_k \text{ and } E_k w_k \text{ agree at } x_j,\ j = 1, 2, \cdots, m.$$

Obviously, in essence, E_k is a matrix of order $m \times m_k$. Let $\mathcal{A}$ and $\mathcal{A}_k$ be the stiff matrices of $A(\cdot, \cdot)$ with respect to V_h^0 and $V_h^{k,0}$. U represents the point parameter value vector of u_h, the solution of (11.1.4). Then (11.1.4) can be written as

$$\mathcal{A}U = F. \tag{11.3.2}$$

By a simple analysis, we know that the matrix form of the projection operator P_k from V_h^0 to $V_h^{k,0}$ is $E_k \mathcal{A}_k^{-1} E_k^T \mathcal{A}$ and Algorithm 11.3.1 is equivalent to

$$U^{n+1} = (I - Q^{-1}\mathcal{A})U^n + Q^{-1}F, \tag{11.3.3}$$

where

$$Q = (\frac{1}{N} \sum_{k=1}^{N} E_k \mathcal{A}_k^{-1} E_k^T)^{-1} \tag{11.3.4}$$

is the induced preconditioner of Algorithm 11.3.1.

11.4 Numerical experiment

Although the experiment is done in the case $N = 2$ (only with two subdomains), we think that for overlap domain decomposition, our numerical results should be true in the case $N > 2$ (with more than two subdomains).

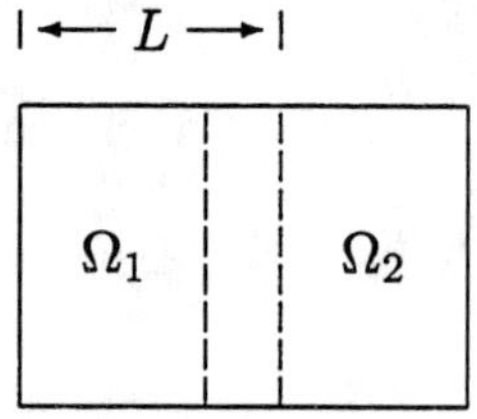

Fig.11.4.1

Decompose $\Omega = (0, 1.5) \times (0, 1)$ into two subdomains: $\Omega_1 = (0, L) \times (0, 1)$ and $\Omega_2 = (1.5 - L, 1.5) \times (0, 1)$, $(L > 0.75)$. See Fig.11.4.1. Ω is triangulated into the fine mesh Ω_h such that each element $e \in \Omega_h$ is an isosceles right triangle with h as its diameter. When $h = 1/4$, there are 48 elements, 117 interpolation points in Ω. When $h = 1/8$, there are 192 elements, 425 interpolation points in Ω. When $h = 1/16$, there are 768 elements, 1617 interpolation points in Ω. The results of applying Algorithm 11.3.1 to the solving of (11.1.4) are listed in Table 11.4.1, Table 11.4.2 and Table 11.4.3, where n represents the number of iterations, $\varepsilon^n = u^n - u_h$ is the error function after n iterations, and

$$E_n = A(\varepsilon^n, \varepsilon^n), \qquad M_n = \max_{x \in \Omega} |\varepsilon^n(x)|,$$
$$\rho_n = \sqrt[n]{E_n / E_0}, \qquad \delta_n = \sqrt[n]{M_n / M_0}.$$

From these three tables, we see that if L is fixed (the overlapness is kept constant), then Algorithm 11.3.1 converges independently of h; if h is fixed (the fine mesh is not changed), then the more overlap the subdomains are (enlarging L), the faster Algorithm 11.3.1 converges.

By [79,80] and the inverse inequality in [26], we can prove that the condition number of the stiff matrix $\mathcal{A}$ is cond($\mathcal{A}$)=$O(h^{-4})$. When h is very small, not only the scale of (11.1.4) is very large, but also the iterative methods, such as the conjuate gradient method (CG), will converge very slowly. For this reason, we employ Q in (11.3.4) as a preconditioner to accelerate the CG for (11.1.4), i.e. we use Algorithm 1.4.1 to solve (11.1.4). At each iteration, it is only needed to calculate

$$Q^{-1} W = \frac{1}{N} \sum_{k=1}^{N} E_k \mathcal{A}_k^{-1} E_k^T$$

which is essentially equivalent to that if $w \in V_h^0$ is determined uniquely by W, then solving the homogeneous boundary value biharmonic equation (the function of $\mathcal{A}_k^{-1}$) with the restrication of w on $V_h^{k,0}$ as the right term (the function of E_k^T), zero–extending it to V_h^0 (the function of E_k) and averaging them, we get $Q^{-1}\mathcal{A}$, the vector form of the average. The numerical results are listed in Table 11.4.4, Table 11.4.5 and Table 11.4.6 to compare CG and PCG used to solve (11.3.2). It is observed that PCG converges in several steps, faster than CG; if the fine mesh is kept constant, then the larger the L, the less steps the PCG, and if L is constant, then the convergent rate is almost independently of h.

Table 11.4.1: ρ_n, δ_n vs. h when $L = 1.25$

	n	5	10	15	20
ρ_n	$h = 0.250$	0.2500	0.2500	0.2500	0.2500
	$h = 0.125$	0.2500	0.2500	0.2500	0.2500
	$h = 0.0625$	0.2500	0.2500	0.2500	0.2500
δ_n	$h = 0.250$	0.5001	0.5001	0.5001	0.5001
	$h = 0.125$	0.5000	0.5000	0.5000	0.5000
	$h = 0.0625$	0.5000	0.5000	0.5000	0.5000

Table 11.4.2: ρ_n, δ_n vs. L when $h = 0.0625$

n		5	10	15	20
ρ_n	$L = 0.875$	0.2505	0.2924	0.4071	0.4861
	$L = 1.000$	0.2499	0.2506	0.2616	0.3000
	$L = 1.125$	0.2498	0.2499	0.2500	0.2500
	$L = 1.250$	0.2500	0.2500	0.2500	0.2500
	$L = 1.375$	0.2497	0.2499	0.2499	0.2499
δ_n	$L = 0.875$	0.5000	0.5111	0.6193	0.6818
	$L = 1.000$	0.5000	0.5000	0.5000	0.5338
	$L = 1.125$	0.5000	0.5000	0.5000	0.5000
	$L = 1.250$	0.5000	0.5000	0.5000	0.5000
	$L = 1.375$	0.5000	0.5000	0.5000	0.5000

Table 11.4.3: When $h = 0.0625$, and E_n reduced from 10^7 to 10^{-11}

L	0.875	1.000	1.125	1.250	1.375
n	162	46	31	30	29
CPU second	181	84	86	95	105

Table 11.4.4: ρ_n, δ_n vs. h for CG

n		10	20	30	40	50	60	70
ρ_n	$h = 0.250$	0.628	0.625	0.618	0.625	0.622	0.618	0.636
	$h = 0.125$	0.742	0.802	0.823	0.815	0.797	0.809	0.814
	$h = 0.0625$	0.860	0.860	0.894	0.900	0.897	0.907	0.895
δ_n	$h = 0.250$	0.845	0.815	0.797	0.803	0.795	0.793	0.802
	$h = 0.125$	0.899	0.907	0.907	0.906	0.893	0.898	0.901
	$h = 0.0625$	0.931	0.948	0.961	0.955	0.950	0.952	0.948

Table 11.4.5: ρ_n, δ_n vs. h when $L = 1.25$ for PCG

	n	1	2	3	4
ρ_n	$h = 0.250$	0.000012	0.000055	0.000169	0.000367
	$h = 0.125$	0.000339	0.000067	0.000197	0.000417
	$h = 0.0625$	0.002450	0.000051	0.000168	0.000518
δ_n	$h = 0.250$	0.002823	0.011649	0.016325	0.024104
	$h = 0.125$	0.004549	0.009340	0.014643	0.022168
	$h = 0.0625$	0.007850	0.008226	0.013711	0.020641

Table 11.4.6: When $h = 0.0625$, and E_n reduced from 10^8 to 10^{-3}

	PCG method					CG
L	0.875	1.000	1.125	1.250	1.375	method
n	6	4	3	3	2	401
CPU second	29	25	23	21	19	921

Bibliography

[1] Adams, R.A., *Sobolev Spaces.* Academic Press, New York, 1975.

[2] Agmon, S., *Lectures on Elliptic Boundary Value Problems.* D. Van Nostrand Company, Inc., Princeton, 1965.

[3] Agoshkov, V.I., Poincaré–Steklov's operates and domain decomposition method in finite dimensional spaces. In [36], 1988.

[4] Argyris, J.H., Fried, I. and Scharpf, D.W., The TUBA family of plate elements for the matrix displacement method. Aero. J. Roy. Aero. Soc., 72(1968), 701–709.

[5] Arnold, D.N. and Brezzi, F., Mixed and nonconforming finite element methods: implementation, postprocessing and error estimates. M^2AM., 19(1985), 7–32.

[6] Babuska, I., The finite element method with Lagrangian multipliers. Numer. Math., 20(1973), 179–192.

[7] Bank, R.E., Dupont, T.F. and Yserentant, H., The hierarchical basis multigrid method. Numer. Math., 52(1988), 427–458.

[8] Bjørstad, P.E. and Widlund, O.B., To overlap or not to overlap: note on a domain decomposition method for elliptic problems. SIAM J. Sci. Stat. Comput., 10(1989), 1053–1061.

[9] Bourgat, J.F., Glowinski, R., Tallec, P.L. and Vidrascu, M., Variational formulation and algorithm for trace operator in domain decomposition calculations. In [23], 1989.

[10] Bramble, J.H. and Pasciak, J.E., A preconditioning technique for indefinite systems resulting from mixed approximations of elliptic problems. Math. Comp., 50(1988), 1–17.

[11] Bramble, J.H., Pasciak, J.E. and Schatz, A.H., An iterative method for elliptic problems on regions partitioned into substructures. Math. Comp., 46(1986), 361–369.

[12] Bramble, J.H., Pasciak, J.E. and Schatz, A.H., The construction of preconditioners for elliptic problems by substructuring. I. Math. Comp., 47(1986), 103–134.

[13] Bramble, J.H., Pasciak, J.E. and Schatz, A.H., The construction of preconditioners for elliptic problems by substructuring. II. Math. Comp., 49(1987), 1–16.

[14] Bramble, J.H., Pasciak, J.E. and Schatz, A.H., The construction of preconditioners for elliptic problems by substructuring. III. Math. Comp., 51(1988), 415–430.

[15] Bramble, J.H., Pasciak, J.E. and Schatz, A.H., The construction of preconditioners for elliptic problems by substructuring. IV. Math. Comp., 53(1989), 1–24.

[16] Bramble, J.H., Pasciak, J.E. and Xu, J., Parallel multilevel preconditioners. Math. Comp., 55(1990), 1–22.

[17] Brandt, A., Multi–level adaptive solution to boundary value problems. Math. Comp., 31(1977), 333–391.

[18] Brenner, S.C., An optimal–order nonconforming multigrid method for the biharmonic equation. SIAM J. Numer. Anal., 26:5(1989), 1124–1138.

[19] Brenner, S.C., A two–level additive Schwarz preconditioner for nonconforming plate elements. Numer. Math., 72(1996), 419–447.

[20] Cai, X., *Some Domain Decomposition Algorithms for Nonselfadjoint Elliptic and Parabolic PDE's.* Ph.D. thesis, Courant Institute, 1989.

[21] Carey, G.F., An analysis of finite element equations and mesh subdivision. Comput. Methods Appl. Mech. Engrg., 9(1976), 165–179.

[22] Chan, T.F., E, W. and Sun, J., Domain decomposition interface preconditioners for fourth order elliptic problems. Appl. Numer. Math., 8(1991), 317–331.

[23] Chan, T.F., Glowinski, R., Périaux, J. and Widlund, O.B., (eds.) *Domain Decomposition Methods.* SIAM Pb., Philadelphia, 1989.

[24] Chu, D., *Some Problems in Domain Decomposition Methods for Elliptic Problems.* Ph.D. thesis, (in Chinese), Tsinghua University, 1991.

[25] Chu, D. and Hu, X., A preconditioned conjugate gradient method with nonoverlap domain decomposition. (in Chinese). Math. Num. Sin., 15:1(1993), 58–68.

[26] Ciarlet, P.G., *The Finite Element Method for Elliptic Problems.* North–Holland, Amsterdam, 1978.

[27] Crouzeix, M. and Raviart, P.A., Conforming and nonconforming finite element methods for solving the stationary Stokes equations. RAIRO Numer. Anal., 7–R3(1973), 33–76.

[28] Dauge, M., *Elliptic Boundary Value Problems on Corner Domains: Smoothness and Asymptotics of Solutions.* Springer–Verlag, New York, 1988. Lecture Notes in Math., Vol. 1341.

[29] Dinh, Q.V., Glowinski, R. and Périaux, J., Solving elliptic problems by decomposition methods with applications. In: *Eliptic Problem Solvers II,* Academic Press, New York, 1982.

[30] Dryja, M., A capacitance matrix method for Dirichlet problem on polygonal region. Numer. Math., 39(1982), 51–64.

[31] Dryja, M. and Proskurowski, W., A capacitance matrix methods using strips with alternating Neumann and Dirichlet boundary conditions. Appl. Numer. Math., 1(1985), 285–298.

[32] Dryja, M. and Widlund, O.B., An additive variant of the Schwarz alternating method in the case of many subregions. Technical Report 339, Courant Institute, 1987.

[33] Ewing, R.E. and Wang, J., Analysis of the Schwarz algorithm for mixed finite element methods. RAIRO Math. Model Numer. Anal., 26:2(1992), 739–756.

[34] Fortin, M. and Soulie, M., A nonconforming piecewise quadratic finite element on triangles. International J. Num. Methods in Engrg., 19(1983), 505–520.

[35] Gilbarg, D. and Trudinger, N.S., *Elliptic Partial Differential Equations of Second Order.* Springer–Verlag, Berlin and New York, 1977.

[36] Glowinski, R., Golub, G.H., Meurant, G.A. and Périaux, J., *Proceedings of First International Symposium on Domain Decomposition Methods for Partial Differential Equations.* SIAM Pb., Philadelphia, 1988.

[37] Glowinski, R., Périaux, J. and Dinh, Q.V., Domain decomposition methods for nonlinear problems in fluid dynamics. Report INRIA 147, 1982.

[38] Glowinski, R., Périaux, J., Shi, Z-C. and Widlund, O., *Domain Decomposition Methods in Sciences and Engineering.* 8th International Conference, Beijing, P.R. China. John Wiley and Sons, Ltd., 1997.

[39] Golub, G.H. and Mayers, D., The use of preconditioning over irregular regions. Lecture at Sixth International Conference on Computing Methods in Applied Sciences and Engineering. Versailles, France, 1983.

[40] Grisvard, P., *Elliptic Problems in Nonsmooth Domain.* Pitman, Boston, 1985.

[41] Gu, J., *Domain Decomposition Methods with Nonconforming Finite Elements.* Ph.D. thesis, (in Chinese), Tsinghua University, 1994.

[42] Gu, J. and Hu, X., On an essential estimate in the analysis of domain decomposition methods. J. Comput. Math., 12:2(1994), 132–137.

[43] Gu, J. and Hu, X., On the overlap domain decomposition methods for elliptic problems in multi–subdomain case. (in Chinese). J. Tsinghua University, 34:6(1994), 50–57.

[44] Gu, J. and Hu, X., On domain decomposition methods in two–subdomain nonoverlap case. Chinese J. Num. Math. & Appl., 17:1(1995), 78–94.

[45] Gu, J. and Hu, X., Overlapping domain decomposition method for nonselfadjoint elliptic problems discretized by Crouzeix–Raviart elements. Chinese J. Num. Math. & Appl., 17:4(1995), 34–42.

[46] Gu, J. and Hu, X., Trace averaging domain decomposition method with nonconforming finite elements. J. Comput. Math., 14:1(1996), 40–53.

[47] Gu, J. and Hu, X., The construction of preconditioners for elliptic problems discretized by nonconforming finite elements via substructuring (I). (in Chinese). Math. Num. Sin., 18:2(1996), 113–128.

[48] Gu, J. and Hu, X., An iterative substructuring method with nonconforming elements. Numerical Mathematics —A Journal of Chinese Universities, 6:1(1997), 77–89.

[49] Gu, J. and Hu, X., Extension theorems for plate elements with applications. Math. Comp., 66(1997), 1375–1388.

[50] Gu, J. and Hu, X., Some estimates with nonconforming finite elements in domain decomposition analysis. J. Comput. Math., 16:1(1998), 63–80.

[51] Gu, J. and Hu, X. and Li, M., Substructuring preconditioners for nonconforming finite element discretizations. Chinese J. Num. Math. & Appl., 19:1(1997), 1–11.

[52] Hackbusch, W., *Multi–Grid Methods and Applications.* Springer–Verlag, New York, 1985.

[53] Han, H., Nonconforming elements in the mixed finite element method. J. Comput. Math., 2:3(1984), 223–233.

[54] Hanisch, M.R., Multigrid preconditioning for the biharmonic Dirichlet problem. SIAM J. Numer. Anal., 30(1993), 184–214.

[55] Jiang, J. and Cheng, X., A nonconforming element like Wilson's for second order problems. (in Chinese). Math. Num. Sin., 14:3(1992), 274–278.

[56] Johnson, C., Crosswind smear and pointwise errors of streamline diffusion finite element methods. Math. Comp., 49(1987), 25–38.

[57] Johnson, C., On the convergence of some mixed finite element methods for plate bending problems. Numer. Math., 21(1973), 43–62.

[58] Keyes, D. and Gropp, W., A comparison of domain decomposition techniques for elliptic partial differential equations. SIAM J. Sci. Stat. Comp. 5(1987), 166–202.

[59] Lascaux, P. and Lesaint, P., Some nonconforming finite elements for the plate bending problem. RAIRO, R–1(1975), 9–53.

[60] Lions, J.L. and Magenes, E., *Non-homogeneous Boundary Value Problems and Applications. I.* Grund. B. 181, Berlin, Springer, 1972.

[61] Lions, P.L., On the Schwarz alternating method I. In [36], 1988.

[62] Lions, P.L., On the Schwarz alternating method II. In [23], 1989.

[63] Lu T., Shih, T.M. and Liem, C.B., Two synchronous parallel algorithms for partial differential equations. J. Comput. Math., 9:1(1991), 74–85.

[64] Mandel, J. and McCormick, S., Iterative solution of elliptic equations with refinement. In [23], 1989.

[65] Marini, L.D. and Quarteroni, A., A relaxation procedure for domain decomposition methods using finite elements. Numer. Math., 55(1989), 575–598.

[66] McCormick, S. and Thomas, J., The fast adaptive composite grid method (FAC) for elliptic boundary value problems. Math. Comp., 46(1986), 439–456.

[67] Morley, L.S.D., The triangular equilibrium problem in the solution of plate bending problems, Aero. Quart., 19(1968), 149–169.

[68] Nečas, J., *Les Méthodes Directes en Théorie des Équations Elliptiques.* Academia, Prague, 1967.

[69] Oden, J.T. and Reddy, J.N., *An Introduction to the Mathematical Theory of Finite Elements.* Wiley, New York, 1976.

[70] Oswald, P., Preconditioners for discretizations of the biharmonic equation by rectangular finite elements. Tech. Report, Friedrich Schiller Universität, Jena, Germany, 1991.

[71] Oswald, P., Hierarchical conforming finite element methods for the biharmonic equation. SIAM J. Numer. Anal., 29(1992), 1610–1625.

[72] Oswald, P., On a hierarchical basis multilevel method with nonconforming $P1$ elements. Numer. Math., 62(1992), 189–212.

[73] Peisker, P., Multilevel algorithms for the biharmonic problem. Numer. Math., 46(1985), 623–634.

[74] Proskurowski, W. and Widlund, O.B., On the numerical solution of Helmholtz's equation by the capacitance matrix method. Math. Comp., 30(1976), 433–468.

[75] Przemieniecki, J.S., Matrix structural analysis of substructures. AIAA J., 1(1963), 138–147.

[76] Schwarz, H.A., Über einige Abbildungsaufgaben. Ges. Math. Abh., 11(1869), 65–83.
[77] Shi, Z., Error estimates of Morley element. Chinese J. Num. Math. & Appl., 12:3(1990), 102–108.
[78] Smith, B. and Widlund, O.B., A domain decomposition algorithm based on a change to a hierarchical basis. SIAM J. Sci. Stat. Comput., 11:6(1990), 1212–1220.
[79] Stummel, F., Basic compactness properties of nonconforming and hybrid finite element spaces. RAIRO Num. Anal., 4:1(1980), 81–115.
[80] Stummel, F., Nonconforming finite element methods and related approximation methods. In: *Proceedings of the China–France Symposium on Finite Element Methods.* Science Press, Beijing, 1983.
[81] Sun, J., Multilevel preconditioners for 4th order problems and domain decomposition methods. In: *Proceedings of 6th International Conference on Domain Decomposition*, Quarteroni, A., et al (eds.), Italy, 1992.
[82] Wang, J. and Yan, N., A parallel domain decomposition procedure for convection–diffusion problems. In [38], 1997.
[83] Widlund, O.B., Iterative methods for elliptic problems on regions partitioned into substructures and the biharmonic Dirichlet problems. In: *Computing Methods in Applied Science and Engineering, VI.* Glowinski, R. and Lions, J.L. (eds.), North–Holland, Amsterdam, 1984.
[84] Widlund, O.B., An extension theorem for finite element spaces with three applications. In: *Numerical Techniques in Continuum Mechanics*, Vol.16, Hackbusch, W. (eds.), Braaunschweig Wiesbaden, 1987.
[85] Widlund, O.B., Preconditioners for spectral and mortar finite element methods. In [38], 1997.
[86] Wilson, E.L. and Taylor, R.L., Incompatible displacement models. In: *Proceedings of the Symposium on Numerical and Computer Methods in Structural Engineering.* University of Illinois, 1971.
[87] Xu, J., *Theory of Multilevel methods.* Ph.D. thesis, Cornell University, 1989.
[88] Yserentant, H., On the multi–level splitting of finite element space. Numer. Math., 49(1986), 379–412.
[89] Zhang, S., *Splitting Parallel Algorithms for Elliptic Equations.* Ph.D. thesis, (in Chinese), Computing Center, Academia Sinica, 1989.
[90] Zhang, X., *Studies in Domain Decomposition: Multilevel Methods and the Biharmonic Dirichlet Problem.* Ph.D. thesis, Courant Institute, 1991.
[91] Zhang, X., Multilevel Schwarz methods for the biharmonic Dirichlet problem. SIAM J. Sci. Comput., 15:3(1994), 621–644.
[92] Zhang, X., Two–level Schwarz methods for the biharmonic problem discretized by conforming C^1 elements. SIAM J. Numer. Anal., 33(1996), 555–570.

Glossary of Symbols

$\Re^n$: Euclidean space of n dimensions.
Ω: domain in $\Re^n$ for integer $n \geq 2$.
Ω_h: fine triangulation of the domain Ω.
Ω_H: coarse triangulation of the domain Ω.
e: typical element in Ω_h (a triangle or a quadrilateral).
Ω_i: the ith subdomain.
h: finite element mesh parameter of Ω_h.
H: finite element mesh parameter of Ω_H.
c: generic positive constant independent of h.
C: generic positive constant independent of h.
$\partial\Omega$: boundary of the domain Ω.
V_h: finite element space associated with Ω_h.
V_h^0: subspace of V_h with zero degrees of freedom at interpolation points of $\partial\Omega$.
V_h^i: subspace of V_h^0 with zero degrees of freedom at interpolation points of $\Omega\backslash\overline{\Omega}_i$.
$V_h^{i,0}$: subspace of V_h^0 with zero degrees of freedom at interpolation points of $\Omega\backslash\Omega_i$.
I_h: conforming interpolation operator.
π_h: finite element interpolation operator associated with V_h.
$\overline{S}$: closure of the set S.
$v|_S$: restriction of the function v to the set S.
$\text{meas}(S)$: measure of the set S.
$\text{supp}\, v$: support of the function v, i.e. $\overline{\{x \in \Re^n : \ v(x) \neq 0\}}$.
$\text{diam}\, \Omega$: diameter of the domain Ω.
$(\cdot,\cdot)$: L^2–inner product.
$a(\cdot,\cdot)$: bilinear form of elliptic problems.
$A(\cdot,\cdot)$: corresponding form of $a(\cdot,\cdot)$ over the triangulation Ω_h.
$A_i(\cdot,\cdot)$: corresponding form of $a(\cdot,\cdot)$ over the triangulation $\Omega_h \cap \Omega_i$.
$\int_S v$: integral of the function v on the set S.
$X + Y = \{x + y : \ x \in X, \ y \in Y\}$.
$X \oplus Y = \{x + y : \ x \in X, \ y \in Y\}$ when $X \cap Y = \{0\}$.
$\mathcal{D}(\Omega) = \{v \in C^\infty(\Omega) : \ \text{supp}\, v \text{ is a compact subset of } \Omega\}$.

$$|v|_{1,\Omega,h} = \Big(\sum_{e\subset\Omega} |v|_{1,e}^2\Big)^{\frac{1}{2}}.$$

$$|v|_{1,\Omega_i,h} = \Big(\sum_{e\subset\Omega_i} |v|_{1,e}^2\Big)^{\frac{1}{2}}.$$